THE BIOLOGY BOOK

WORKBOOK

UNITS **1** **2**

Pam Borger
Kelli Grant
Louise Munro
Jane Wright

The Biology Book Units 1 & 2
1st Edition
Pam Borger
Kelli Grant
Louise Munro
Jane Wright

Contributing authors: Daniel Avano, Andrea Blunden, Sue Farr, Sarah Jones and Katrina Walker

Publishing editor: Rachel Ford
Project editors: Nadine Anderson-Conklin and Siobhan Moran
Copy editor: Catherine Greenwood
Proofreader: Siobhan Moran
Permissions researcher: Wendy Duncan
Text design: Leigh Ashforth (Watershed Design)
Cover design: Chris Starr (MakeWork)
Cover image: iStock.com/defun
Project design: Aisling Gallagher
Production controller: Erin Dowling
Typeset by: MPS Limited

Any URLs contained in this publication were checked for currency during the production process. Note, however, that the publisher cannot vouch for the ongoing currency of URLs.

© 2018 Cengage Learning Australia Pty Limited

For product information and technology assistance,
in Australia call **1300 790 853**;
in New Zealand call **0800 449 725**

For permission to use material from this text or product, please email **aust.permissions@cengage.com**

ISBN 978 0 17 041166 0

Cengage Learning Australia
Level 7, 80 Dorcas Street
South Melbourne, Victoria Australia 3205

Cengage Learning New Zealand
Unit 4B Rosedale Office Park
331 Rosedale Road, Albany, North Shore 0632, NZ

For learning solutions, visit **cengage.com.au**

Printed in China by 1010 Printing International Limited.
2 3 4 5 6 7 23

CONTENTS

UNIT 1 » CELLS AND MULTICELLULAR ORGANISMS   1

TOPIC 1: CELLS AS THE BASIS OF LIFE   2

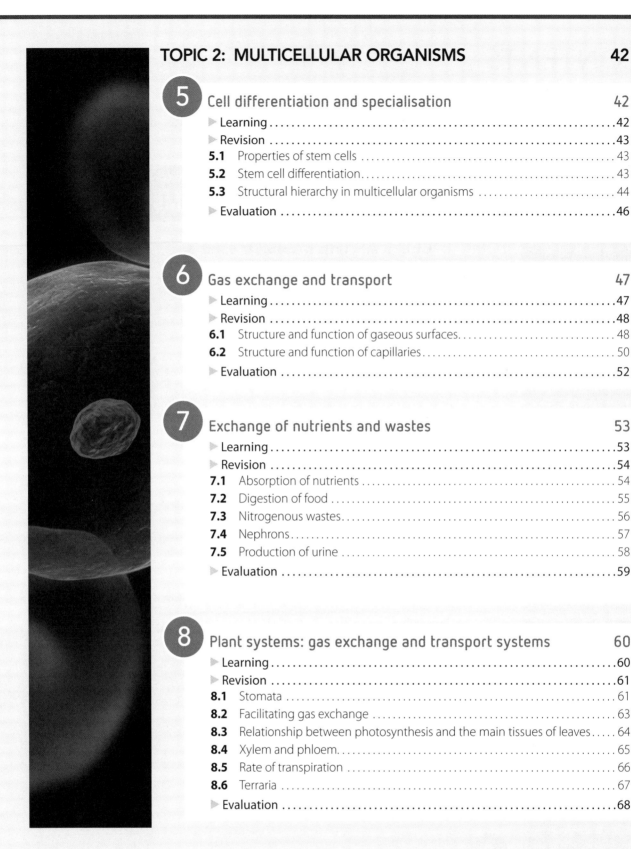

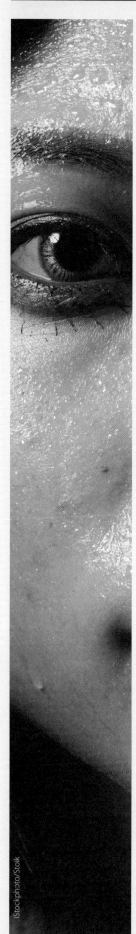

iStockphoto/Stolk

HOW TO USE THIS BOOK

Learning

The learning section is a summary of the key knowledge and skills. This summary can be used to create mind maps, to write short summaries and as a check list.

Revision

The revision section is a series of structured activities to help consolidate the knowledge and skills acquired in class.

Evaluation

The evaluation section is in the style of a practice exam to test and evaluate the acquisition of knowledge and skills.

Practice exam

A tear-out exam helps to facilitate preparing and practising for external exams.

ABOUT THE AUTHORS

Pam Borger

Pam is a highly experienced biology teacher and author. Among Pam's writing credits is the successful *Jumpstart Biology*. Pam led the team on *Nelson Biology for the Australian Curriculum* and *The Biology Books*.

Kelli Grant

Kelli has a breadth of biology teaching experience across the high school, TAFE and university sectors. Kelli has written for *Nelson Biology for the Australian Curriculum* and *The Biology Books*.

Louise Munro

Louise is a highly experienced teacher of senior biology and chemistry. Louise has written for *Nelson Biology for the Australian Curriculum* and *The Biology Books*.

Jane Wright

Jane is an experienced biology teacher and writer. Jane has written for *Nelson Biology for the Australian Curriculum* and *The Biology Books*.

Some of the material in *The Biology Book Units 1 & 2* has been taken from or adapted from the following: *Nelson Biology for the Australian Curriculum Units 3 & 4* NelsonNet material written by: Andrea Blunden, Sarah Jones, Sue Farr and Daniel Avano.

Nelson Biology for the Australian Curriculum Units 1 & 2 NelsonNet material written by: Andrea Blunden, Pam Borger and Katrina Walker.

SYLLABUS REFERENCE GRID

UNITS AND TOPICS	THE BIOLOGY BOOK UNITS 1 & 2
UNIT 1: CELLS AND MULTICELLULAR ORGANISMS	
Topic 1: Cells as the basis of life	
Cell membrane	Chapter 1
Prokaryotic and eukaryotic cells	Chapter 2
Internal membranes and enzymes	Chapter 3
Energy and metabolism	Chapter 4
Topic 2: Multicellular organisms	
Cell differentiation and specialisation	Chapter 5
Gas exchange and transport	Chapter 6
Exchange of nutrients and wastes	Chapter 7
Plant systems: gas exchange and transport systems	Chapter 8
UNIT 2: MAINTAINING THE INTERNAL ENVIRONMENT	
Topic 1: Homeostasis	
Homeostasis	Chapter 9
Neural homeostatic control pathways	Chapter 10
Hormonal homeostatic control pathways	Chapter 11
Thermoregulation	Chapter 12
Osmoregulation	Chapter 13
Topic 2: Infectious diseases	
Infectious disease	Chapter 14
Immune response and defence against disease	Chapter 15
Transmission and spread of disease	Chapter 16

9780170411660

UNIT ONE

CELLS AND MULTICELLULAR ORGANISMS

- Topic 1: Cells as the basis of life

- Topic 2: Multicellular organisms

9780170411660

1 Cell membrane

LEARNING

Summary

- The cell membrane forms a barrier between the internal and external environments of cells.
- The structure of the cell membrane is described by the fluid mosaic phospholipid bilayer model.
- The fluid mosaic model states that all membranes are composed of a phospholipid bilayer into which proteins are embedded.
- The plasma membrane regulates the internal cellular environment by allowing some molecules to pass across it but not others.
- Molecules on the surface of the plasma membrane allow cells to recognise each other and respond to chemical messengers such as hormones.
- Simple diffusion, facilitated diffusion and osmosis are ways that molecules can cross membranes by passive transport that does not require energy.
- Active transport uses energy to move substances across membranes, up their concentration gradient.
- Endocytosis, which includes phagocytosis, is a type of active transport in which substances move into cells in membrane-bound vesicles.
- The physical and chemical nature of a substance determines the way in which it will be transported across membranes.
- The surface-area-to-volume ratio of cells affects the adequate supply of nutrients and removal of wastes, and therefore limits cell size.
- The greater the concentration gradient of a substance across a membrane, the faster it will diffuse.

1.1 Structure and function of cell membranes

Because the cell membrane could not be seen under the light microscope, scientists initially took an indirect approach to understanding it by making models using the membrane's physical and chemical properties. In the mid-1890s, Charles Ernest Overton concluded that because a lipid-soluble substance such as chloroform readily enters cells, the cell membrane was composed of lipids. In 1925, further research indicated that the lipids were arranged in a bilayer.

A decade later, to account for the observation that around half of the mass of plasma membranes was protein, Davson and Danielli proposed a model in which the lipid bilayer was coated on either side with a layer of proteins. Although direct observation of membranes with the newly invented electron microscope in the 1950s appeared to confirm this model, not all experimental observations fitted it neatly. For example, the model did not explain the fluidity of the membranes observed in living cells. In 1972, Singer and Nicolson proposed the fluid mosaic model, which was able to explain all the physical and chemical properties of membranes known at that time.

QUESTIONS

1 In the space below, draw and label part of a cell membrane. Ensure that you include and label the phospholipid bilayer, a cholesterol molecule and a transport, receptor, recognition and adhesion protein.

2 Explain why Singer and Nicolson used the words 'fluid', 'mosaic' and 'model' to describe their ideas.

_____

_____

_____

_____

3 In relation to membrane phospholipids, describe what is meant by 'hydrophilic' and 'hydrophobic' and explain how these features give rise to the phospholipid bilayer.

_____

_____

4 Use the information given to find evidence to support or refute the statement 'New discoveries in science often depend on advances in technology'.

_____

_____

_____

_____

5 Complete the table by writing the function of each structure in the right-hand column.

STRUCTURE	FUNCTION
Receptor molecule	
Carrier molecule	
Protein channel	
Recognition molecule	
Adhesion molecule	

1.2 | Curing cancer: changes to the cell membrane mark cancer cells for death

Dr Barbara Sanderson from the Flinders Medical Centre in Adelaide, South Australia, has shown that the milky venom from a sea anemone kills human lung cancer cells. The venom induces apoptosis, a particular type of cell death that cancer cells in the body can usually avoid.

Cell membrane receptors on cancer cells are thought to respond to the sea anemone venom by relaying messages to 'death receptors' inside the cell. This activates a group of 'protein-eating' enzymes that change the structure and composition of the cell membrane. The cell membrane then becomes more permeable to small molecules. Other venom enzymes disrupt the phospholipid asymmetry by moving specific molecules to the outer side of the lipid bilayer. This marks the cell for death. The molecules act like 'eat me' signals and the cells are engulfed by roaming white blood cells.

QUESTIONS

1 Name and describe the process used by white blood cells to engulf cells undergoing apoptosis.

_____

_____

_____

_____

2 Compare and contrast active and passive movement across membranes and classify the process described in Question **1** as one of these.

_____

_____

_____

3 Use your knowledge of membrane structure to predict what changes might occur to the cell membrane during apoptosis to make it more permeable to small molecules.

_____

_____

4 Name the type of molecule that forms cell membrane receptors on cancer cells.

_____

5 Use the information given to predict what is meant by saying the cell membrane receptors and the sea urchin venom have complementary shapes.

_____

_____

1.3 Osmosis

Osmosis is the passive diffusion of water across a differentially permeable membrane. It significantly influences both plant and animal cells.

QUESTIONS

1 Define:

a isotonic

_____

b hypertonic

_____

c hypotonic

_____

2 In hospital, patients may be connected to an intravenous drip that adds fluid to their blood plasma. Figure 1.3.1 shows why it is important that the fluid in the drip has a solute concentration equal to blood plasma.

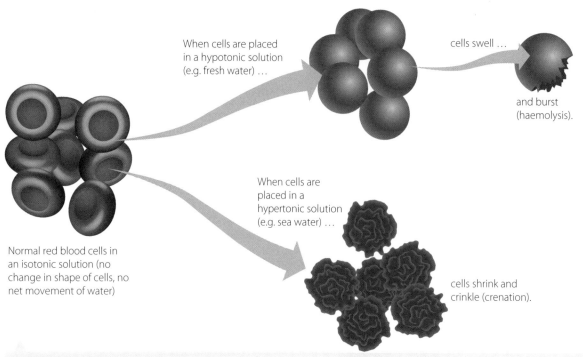

When cells are placed in a hypotonic solution (e.g. fresh water) …

cells swell …

and burst (haemolysis).

When cells are placed in a hypertonic solution (e.g. sea water) …

cells shrink and crinkle (crenation).

Normal red blood cells in an isotonic solution (no change in shape of cells, no net movement of water)

FIGURE 1.3.1 Damaged blood cells. Human red blood cells swell or shrink in solutions of varying solute concentrations.

a Predict what will happen if the intravenous drip contained very salty water and describe the water movement that brings about this change.

_____

_____

b Use Figure 1.3.1 to explain why soaking a blood-stained piece of clothing in cold water will remove the stain.

_____

_____

_____

c Plant tissue, like lettuce, can be soaked in water to freshen it up. Compare and contrast the structure of plant and animal cells to explain their different reactions to soaking in water.

_____

_____

_____

_____

9780170411660

1.4 Surface-area-to-volume ratio

An important concept related to cell size is the surface-area-to-volume ratio (SA:V), which is expressed as a ratio (e.g. 3:1) or a number (e.g. 3). The uptake of materials from the external environment into a cell occurs through its cell membrane. These materials are then used to fuel the chemical reactions that occur throughout the volume of the cytoplasm. For a cell to be able to supply the cytoplasm with its metabolic requirements and remove wastes, it needs a large surface-area-to-volume (SA:V) ratio.

Changes in the shape and size of a cell can change the SA:V ratio.

QUESTIONS

1 Calculate the SA:V ratio of the following cubes.

   a $5\,cm \times 5\,cm \times 5\,cm$

   _____

   b $0.3\,cm \times 0.3\,cm \times 0.3\,cm$

   _____

2 **a** Complete the table by writing in the values for SA, V and SA:V for each cube.

CUBE (mm)	SA (mm^2)	V (mm^3)	SA:V
0.5			
1			
2			
4			

   b What trend is shown by the ratios?

   _____

   c Explain the importance of a high SA:V ratio in terms of cellular function.

   _____

   _____

   _____

   _____

   _____

   _____

   _____

3 a Complete the table by writing in the values for SA, V and SA:V for the rectangular objects.

RECTANGLE	LENGTH (mm)	WIDTH (mm)	HEIGHT (mm)	SA (mm²)	V (mm³)	SA:V
1	32	16	1			
2	16	16	2			
3	8	16	4			
4	8	8	8			

b What happens to the surface area and volume as the height increases?

_____

c What happens to the SA:V ratio as height increases?

_____

d What dimension of the rectangular object would allow the highest rate of diffusion?

_____

e Explain why leaves are long and flat.

_____

1.5 | Second-hand data analysis: osmosis in potatoes

A group of students completed the following practical investigation.

AIM

To demonstrate the process of osmosis in plant tissue

PROCEDURE

1 Cut three cubes of potato, each $3\,cm \times 3\,cm \times 3\,cm$.

2 Cut a well in each cube.

3 Place one cube in a beaker of water and boil it for 3 minutes.

4 Place all three cubes in a dish and add water until it is roughly level with the bottom of the wells.

5 Place a teaspoon of honey into the well of one unboiled cube and the boiled cube. Leave the third empty.

6 Leave for several hours then observe any changes.

RESULTS

The students' results are summarised in Table 1.5.1.

TABLE 1.5.1 The results obtained by group of students following the above method

	UNBOILED WITHOUT HONEY	UNBOILED WITH HONEY	BOILED WITH HONEY
CHANGES IN THE WELLS	No change	Fluid level in the well rose significantly	Fluid level in the well rose a little
OTHER OBSERVATIONS	Surface of potato went brown	Surface of potato went brown	Surface of potato unchanged

9780170411660

DISCUSSION

1   Identify the independent variable and the control in this practical.

_____

2   Explain in terms of osmosis, the changes in the wells shown in Table 1.5.1.

_____

_____

_____

_____

3   Predict the results that would be obtained from other types of vegetables. Give reasons for your predictions.

_____

_____

_____

_____

_____

1 Moss plants are highly salt tolerant. This is because moss cells are able to maintain a low internal sodium concentration even when suspended in salty water. Which of the following mechanisms would explain this observation?

   A The removal of sodium ions by active transport

   B The uptake of sodium ions by diffusion

   C The uptake of water by osmosis

   D The removal of water by active transport

2 GLUT4 glucose transporters carry glucose into muscle cells. When blood insulin concentrations are low, GLUT4 glucose transporters are present in the membranes of cytoplasmic vesicles, where they are unable to transport glucose. Binding of insulin to receptors on the cell surface leads to rapid fusion of the vesicles with the plasma membrane and insertion of the glucose transporters, enabling the cell to efficiently take up glucose. When blood levels of insulin decrease and insulin receptors are no longer occupied, the glucose transporters are recycled back into the cytoplasm. Recent studies of individuals with type 2 diabetes have shown that although insulin is unable to stimulate translocation of GLUT4 to the plasma membrane, a bout of exercise does.

   Choose the best option.

   A The facilitated diffusion of glucose by GLUT4 moves glucose up its concentration gradient into cells.

   B The presence of insulin stimulates endocytosis of the GLUT4 cytoplasmic vesicles; its absence causes exocytosis.

   C Exercise stimulates the fusion of vesicles containing GLUT4 glucose transporters with the plasma membrane.

   D The membrane recognition proteins to which insulin binds are called glycoproteins because they are a protein combined with a sugar molecule.

3 Sugar and water are mixed together.

   a Name the solvent.

   _____

   b Name the solute.

   _____

4 Two solutions are separated by a differentially permeable membrane. The solution on the left of the membrane is 2% salt and the solution on the right of the membrane is 5% salt.

   a Indicate which solution is dilute and which is concentrated.

   _____

   b Predict the direction of the net movement of water molecules and justify your answer.

   _____

   _____

   _____

9780170411660

5 Describe the role of cholesterol in the cell membrane.

_____

_____

_____

_____

6 Calculate the surface-area-to-volume (SA:V) ratio of a cube $2\,cm \times 2\,cm \times 2\,cm$.

7 The concentration of different ions inside and outside the cell was measured and is shown in Table 1.6.1.

TABLE 1.6.1 Comparison of the concentration of various ions inside and outside the cell

TYPE OF ION	EXTRACELLULAR CONCENTRATION (mmol/L)	INTRACELLULAR CONCENTRATION (mmol/L)
Sodium	145	15
Potassium	4.5	120
Chloride	116	20
Calcium	1.2	10

a Use the information in Table 1.6.1 to name one ion that is transported out of the cell by active transport.

_____

b Name the energy source that powers this movement.

_____

c Describe the process by which this ion moves across the membrane.

_____

_____

_____

_____

2 Prokaryotic and eukaryotic cells

LEARNING

Summary

▶ Living things use the energy in food to fuel their activities and they are able to remove wastes.

▶ Autotrophs use energy from the Sun to produce their food.

▶ Heterotrophs rely on autotrophs for food to meet their energy needs.

▶ Carbohydrates, lipids, proteins and nucleic acids are the main macromolecules of living things.

▶ Cells assemble macromolecules from small organic compounds, including simple sugars, fatty acids and glycerol, amino acids and nucleotides.

▶ The development of different kinds of microscopes has advanced our understanding of cell structure.

▶ The simplest type of cell is a prokaryote. Prokaryotes exist as single cells containing a circular chromosome and ribosomes.

▶ The endosymbiotic theory proposes that eukaryote cells were formed when a bacterial cell was ingested by another primitive prokaryotic cell.

▶ Eukaryotic cells are complex cells containing membrane-bound compartments with specific metabolic functions.

▶ Chloroplasts, containing the pigment chlorophyll, are organelles in eukaryotic cells that use the energy in sunlight to convert carbon dioxide and water to glucose and oxygen.

▶ Mitochondria are the organelles in eukaryotic cells where aerobic cellular respiration takes place.

▶ Cellular respiration is the series of chemical reactions that break down glucose and use oxygen to produce carbon dioxide and water. The energy released by this process is used to build up the energy storage molecule ATP.

▶ Proteins are synthesised by the rough endoplasmic reticulum.

▶ Lipids, carbohydrates and steroids are synthesised by the smooth endoplasmic reticulum.

▶ Plastids synthesise pigments, tannins and polyphenols.

▶ Lysosomes remove cellular products and wastes.

2.1 Fundamentals of cell function

CLOZE ACTIVITIES

Fill in the missing words in these cloze activities.

1 Energy:

All cells require _____. Autotrophic organisms make their own energy-containing food through _____. _____ are organisms that must consume others to gain energy. _____ is the metabolic pathway that breaks down _____ to provide energy to the cell.

2 Evolution of eukaryotes:

The initial _____ appeared approximately 2 billion years ago. The _____ theory proposes that the first eukaryote cells were created when a cell was _____ by another primitive _____. The engulfing of a smaller bacterial cell that survives in the host is known as _____. It is thought that _____ and _____ were created in the same way, forming a symbiotic relationship with the host cell. Both chloroplasts and _____ can only arise from chloroplasts and _____ respectively and both have a _____ membrane and their own _____ (genetic material).

3 Photosynthesis:

_____ is the chemical reaction that harvests the Sun's energy and converts it into useable energy. The green pigment _____, found in the _____ of plant cells, is able to harvest the Sun's energy and make it available for use in photosynthesis. _____ are oval-shaped membrane-_____ organelles. Photosynthesis is a series of chemical reactions that produces the energy-rich molecule _____ and the gas _____.

4 Cellular respiration:

_____ is the chemical reaction whereby _____ is broken down into _____ and _____, providing energy to the cell. This reaction occurs in an organelle called the _____. Like _____, mitochondria have a double membrane and their own _____ (genetic material).

2.2 Organelles and macromolecules in cells

QUESTIONS

1 Name the organelle that best matches each of the following statements.

   a Cytoplasmic organelles that contain digestive enzymes

   _____

   b Organelles within the cytoplasm that are the site of aerobic cellular respiration releasing energy for the cell

   _____

   c Membranous site of protein synthesis

   _____

   d A plastid that produces and stores coloured pigments

   _____

   e The organelle in a eukaryotic cell containing DNA whose function is to coordinate cell activities

   _____

   f Endoplasmic reticulum with no ribosomes attached

   _____

   g Colourless plastid not containing pigments

   _____

   h Double membrane bound organelles present in all living plant cells.

   _____

2 Explain the advantage of eukaryotic cells possessing different types of membrane-bound organelles.

   _____

   _____

3 Certain cells have densely packed mitochondria. Predict the function of these cells and explain your reasoning.

   _____

   _____

4 Distinguish between a monomer and a polymer.

   _____

   _____

9780170411660

5 Complete the table by writing a description of the role of each substance in the right-hand column.

SUBSTANCE	ROLE
Urea	
Starch	
Uric acid	
Cellulose	
Glycogen	
Ammonia	
Glycerol	
Steroid	

2.3 | Microscopes, magnification and size

Advances in technology over the last decades have allowed scientists to see and understand the microscopic world.

Centimetres, millimetres and even micrometres or microns (μm) are often too large to measure cells and their contents. A more appropriate measure is nanometres.

$$1 \text{ metre (m)} = 10^2 \text{ centimetres (cm)}$$
$$= 10^3 \text{ millimetres (mm)}$$
$$= 10^6 \text{ micrometres or microns (μm)}$$
$$= 10^9 \text{ nanometres (nm)}$$

QUESTIONS

1 Fill in the missing words in this cloze activity about microscopes.

_____: light rays from a lamp pass through a thin specimen and then through two glass lenses, the _____ and _____ (eyepiece) lenses to our eyes.

_____: much greater _____ and _____ because it uses an _____ beam instead of light to pass through the specimen. It is focused using _____ instead of glass lenses.

Scanning electron microscope (SEM): solid specimens are bombarded with a beam of electrons, providing a lower _____ three-dimensional _____ view of the specimen.

2 When tissue is viewed under a light microscope and then under a transmission electron microscope, which of the following will increase?

A Number of cells visible

B Clarity of the image

C Diameter of field of view

D Three-dimensional depth of the image

3 Using the resources you have available, complete the following.

a Estimate the thickness of a human hair in μm and mm.

_____

b Compare the size of a red blood cell and a virus.

_____

c Calculate how many glucose molecules could fit across the diameter of a red blood cell.

_____

d Comment on the validity of this statement and give your reasons: Many objects can be viewed under both the electron microscope and the light microscope.

_____

_____

9780170411660

4 Circle the correct response: true or false.

a The transmission electron microscope (TEM) shows the surface features of cells because electrons are reflected from the surface layers of the specimen. True/False

b The TEM has greater resolving power than the light microscope because it uses a beam if electrons and electromagnets, instead of a beam of light and glass lenses. True/False

c Chloroplasts can be seen moving around inside a plant cell using the light microscope because living cells can be viewed in a light microscope. True/False

d A three-dimensional picture of a plant cell vacuole can be obtained using a SEM because electrons pass through the specimen. True/False

2.4 Using the light microscope to measure cells

The light microscope can be used to view different cells; for example, red blood cells and human hairs. It can also be used as a measuring device, as described in the following activity.

QUESTIONS

1 Fill in the missing words in this cloze activity.

_____: is calculated by multiplying together the magnifications of the objective and ocular (eyepiece) lenses.

Oil immersion lens: a drop of _____ between a $\times 100$ objective lens and the slide increases the _____ and _____ possible when viewing very small cells such as bacteria.

Field diameter: is the _____ of the field of view, the circle of light, seen down the microscope.
A _____ or mini-grid can be used to measure the _____ of view of a microscope at low power. As magnification _____, the field of view _____. To find the field diameters at higher magnifications, divide the _____ diameter at low power by the change in total _____.

Cell size: If the diameter of the field of view is known, the size of a cell can be estimated as follows:

$$\text{Cell size} = \frac{\text{_____ of field of view}}{\text{_____ of cells across field}}$$

2 Complete the table by writing in the magnifications and field diameters. The field diameter using a $\times 10$ ocular and $\times 10$ objective was measured to be 1.25 mm, which is 1250 μm.

POWER	LENS COMBINATION		TOTAL MAGNIFICATION	FIELD DIAMETERS	
	OCULAR LENS	OBJECTIVE LENS		MILLIMETRES (mm)	MICROMETRES (μm)
Low power	$\times 10$	$\times 10$	$\times 100$	1.25	1250
High power	$\times 5$	$\times 40$			
Oil immersion	$\times 10$	$\times 100$			

3 Five similar cells fit across the field at a total magnification of $\times 100$.

a Calculate the diameter of one cell, in mm and μm.

_____

b Calculate how many of these cells would be seen with the $\times 10$ ocular and oil immersion lens.

_____

1 Cells contain many important macromolecules, including:

   A lipids that form structural components of membranes

   B carbohydrates that increase the rate of reactions

   C nucleotides that are made up of chains of amino acids

   D proteins that are produced in the mitochondria.

2 The amount and function of smooth endoplasmic reticulum depends on the type of cell where it is located. The smooth endoplasmic reticulum does *not*:

   A detoxify drugs

   B manufacture proteins

   C store calcium ions

   D synthesise lipids.

3 Mitochondria and ribosomes are:

   A both membrane-bound organelles

   B only found in eukaryotic cells

   C only found in prokaryotic cells

   D found in liver cells.

4 A student estimated the diameter of a cell to be 8.6 μm. Convert this to millimetres (mm).

5 Describe two ways in which a prokaryote and a eukaryote cell are different and two ways in which they are the same.

_____

_____

_____

_____

6 Describe the function of lysosomes.

_____

_____

_____

_____

_____

7 Name the process in which eukaryote cells were formed when a bacterial cell was ingested by another primitive prokaryotic cell.

_____

8 Distinguish between heterotrophs and autotrophs.

_____

_____

_____

_____

_____

3 Internal membranes and enzymes

LEARNING

Summary

▶ Having membrane-bound organelles creates specialised environments for specific functions.

▶ Compartmentalising allows a large number of cellular activities to occur at the same time in a very limited space and under different conditions.

▶ Control of specific biochemical reactions can be achieved through arrangement of internal membranes.

▶ The internal surface area of mitochondria is large because of the folding and stacking of their internal membranes.

▶ Each step of metabolic reactions that occur in cells is controlled and regulated by enzymes.

▶ Intracellular enzymes speed up and control metabolic reactions inside cells. Extracellular enzymes are produced by cells but function outside the cell.

▶ Chemical reactions in cells occur in a series of regulated steps called biochemical pathways. The product of one step becomes the reactant for the next step.

▶ An enzyme has a precise place on its surface with a specific shape to which a specific substrate can become attached. This is called the active site.

▶ The induced-fit and lock-and-key models describe enzyme action.

▶ Enzymes speed up reactions by reducing the activation energy required to catalyse a reaction.

▶ Enzymes are not destroyed or altered by reactions; they can be reused.

▶ Enzymes can work in either direction of a metabolic reaction.

▶ Enzymes work best in a limited range of temperature and pH.

▶ Enzymes are denatured and cannot function again when temperatures become too high. When temperatures decrease too much, the rate of reaction slows down but the enzyme is not destroyed.

▶ The amount of substrate or enzyme present in a reaction mix can limit the amount of product produced.

▶ Competitive inhibitors bind to an enzyme's active site; non-competitive inhibitors bind to other sites in the enzyme.

▶ Enzyme activity is affected by cofactors and coenzymes.

9780170411660

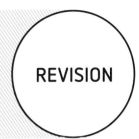

3.1 | Internal membranes in cells

Compartmentalising a cell by having membrane-bound organelles creates specialised environments for specific functions. This enables a large number of activities to occur at the same time in a very limited space and under different conditions. Some organelles, such as mitochondria, increase their internal surface area by the folding and stacking of internal membranes.

Your task is to make a model to illustrate the advantages of internal membranes to cellular function.

INSTRUCTIONS

1   Take an empty matchbox, for instance Redheads© safety matches or a box of similar size. Measure the length, width and height of the box. Calculate the surface area by using the formula:

$$SA = (2 \times length \times width) + (2 \times length \times height) + (2 \times width \times height)$$

Insert your measurements below.

Length of box = _____ cm

Width of box = _____ cm

Height of box = _____ cm

Surface area: (2 × _____ × _____) + (2 × _____ × _____)

+ (2 × _____ × _____) = _____ cm^2

2   Take an A4 piece of paper. Measure its length and width. Calculate the surface area by using the formula:

$$SA = length \times width$$

Insert your measurements below.

Length of A4 paper = _____ cm

Width of A4 paper = _____ cm

Surface area: _____ × _____

= _____ cm^2

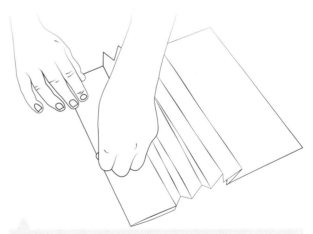

3   Fold the A4 piece of paper into pleats (see Figure 3.1.1). Fold both lengthwise and height wise until the paper fits into the matchbox.

FIGURE 3.1.1 Folding a sheet of A4 paper into pleats

QUESTIONS

1   Describe the difference between the surface area of the empty box and the box and paper model.

_____

_____

2 By referring to your textbook or another suitable source, draw a diagram of a mitochondrion. Label the outer membrane, inner membrane, intermembrane space and matrix.

3 Compare the box and paper model with your sketch of a mitochondrion. Describe the similarities and differences.

_____

_____

_____

_____

4 Consider an empty box containing five matches. Imagine each match as a chemical reaction. Now consider the box and paper model containing five matches in different places (Figure 3.1.2).

Using these two situations, suggest how a large number of chemical reactions can occur at the same time in a very limited space and under different conditions.

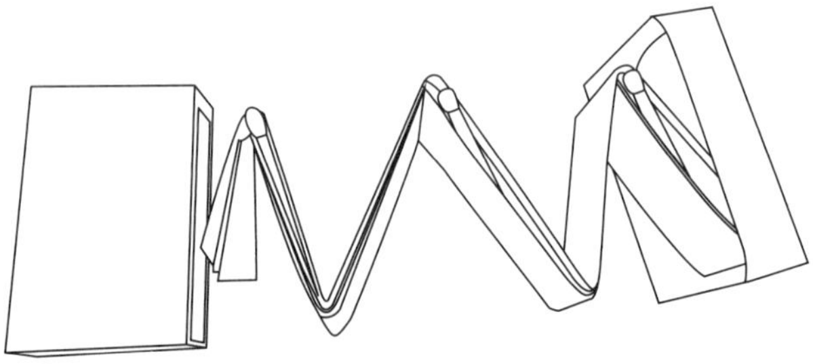

FIGURE 3.1.2 A model to investigate compartmentalisation in mitochondria

_____

_____

_____

_____

5 Reflect on the usefulness of constructing and using models in biology.

_____

_____

_____

_____

9780170411660

3.2 Enzyme action

Enzymes are nature's catalysts. They are present in all metabolic pathways, they are highly specialised and a number of factors affect their function. Enzymes can either break down a substrate or bond two substrates together.

QUESTIONS

1 Define:

a metabolism

_____

b catalyst

_____

c activation energy

_____

2 Draw an annotated diagram of the lock-and-key model. Use the following terms in your annotation.

> enzyme–substrate complex   enzyme   substrate   products   active site

3 Explain the induced-fit model.

_____

_____

_____

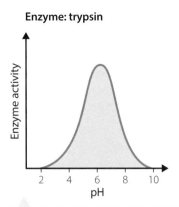

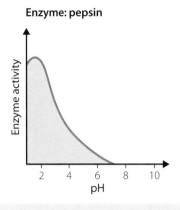

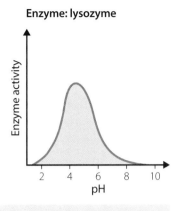

FIGURE 3.2.1 Enzyme activity and pH

4 a Using the graphs in Figure 3.2.1, state the optimum pH for each of the enzymes.

_____

b Pepsin is found in the stomach where it breaks down complex proteins into long peptide chains. Trypsin is found in the small intestine and continues the digestion of proteins. Trypsin breaks down the long peptide chains into short ones. Explain why pepsin no longer functions when it enters into the small intestine.

_____

_____

_____

5 Identify and discuss the function of a non-competitive inhibitor.

_____

_____

_____

_____

_____

_____

9780170411660

3.3 | Properties of enzymes

QUESTIONS

1 How does an 'active site' form?

_____

_____

_____

2 Diastase is an enzyme found in malt that speeds up the breakdown of polysaccharide starch to maltose and then to the glucose monomer at the time of germination (Figure 3.3.1).

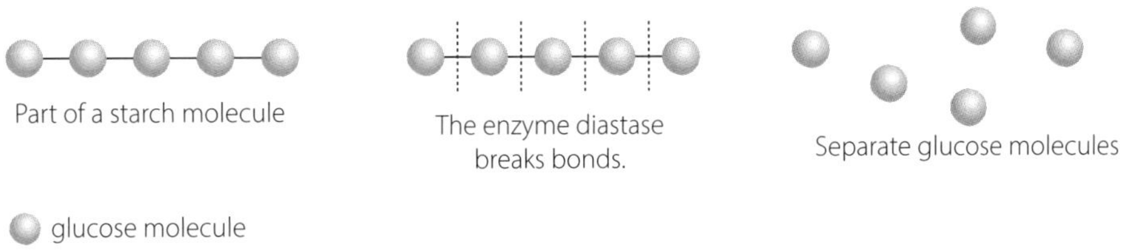

Part of a starch molecule

The enzyme diastase breaks bonds.

Separate glucose molecules

FIGURE 3.3.1 A diagram of the breakdown of starch to glucose

glucose molecule

a Which substance is the substrate in this reaction?

_____

b Which substance is the end product in this reaction?

_____

c Draw a diagram of what the enzyme–substrate complex might look like.

3 Explain why a change in an enzyme's shape causes it to denature.

_____

_____

4 Explain why a doctor would be concerned if their patient developed a temperature of over 42°C.

_____

_____

5 Use Figure 3.3.2 to answer the questions that follow.

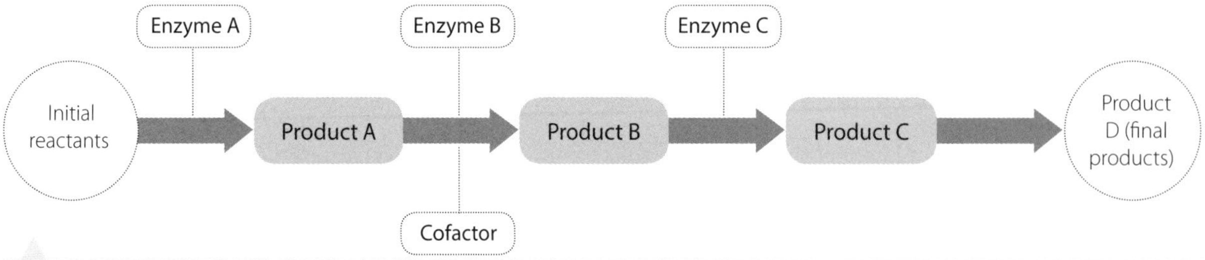

FIGURE 3.3.2 A biochemical pathway

    a If enzyme A in the chain was faulty, name the:

      i product that would be in excess _____

      ii products that would be missing. _____

    b Describe the effect of a faulty version of the cofactor.

    _____

    _____

    c Product A has not been produced in an individual's metabolic pathway. Suggest a possible reason.

    _____

    _____

    d Identify which enzymes could be faulty if product D was not produced.

    _____

    _____

    e If you suspect enzyme B to be faulty, suggest what substances you could test for. Discuss what you would expect to find.

    _____

    _____

6 List four properties of enzymes.

_____

_____

_____

_____

3.4 | Factors affecting enzyme action

QUESTIONS

1 The pH of human blood and body fluids (excluding the gastric juices) is 6.8–7.0. Explain why maintaining this level of pH is important.

_____

_____

_____

2 Figure 3.4.1 shows the effect of substrate concentration on the rate of an enzyme-controlled reaction. The temperature was kept constant.

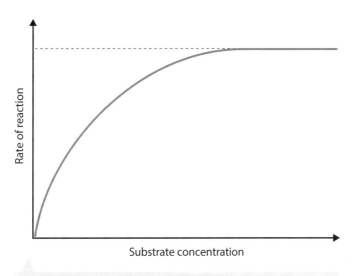

FIGURE 3.4.1 The effect of increases in substrate concentration on the rate of an enzyme-controlled reaction

a Refer to the 'active site' to explain why the rate of reaction levels out although the substrate concentration increases.

_____

_____

_____

b Draw another trend line on Figure 3.4.1 to show what would happen if you doubled the amount of enzyme present.

3 Complete the graph in Figure 3.4.2 by drawing a line to generalise the relationship between reaction rate and cofactor/coenzyme concentration.

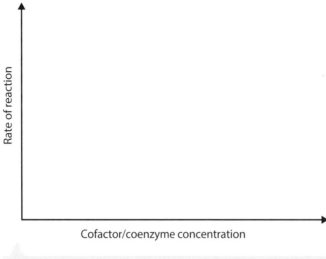

FIGURE 3.4.2

4 Heavy metals such as mercury bind to enzyme active sites. Explain why it is dangerous to accumulate these types of heavy metals in your body.

_____

_____

_____

_____

1 Organisms produce hydrogen peroxide (H_2O_2), a by-product of metabolism that is toxic to cells. Catalase speeds up the reaction:

$$2H_2O_2 \rightarrow 2H_2O + O_2$$

In this reaction, identify the:

a substrate _____

b enzyme _____

c product _____

2 Name the precise place on the surface of an enzyme to which a substrate can become attached.

_____

3 Prokaryotic cells such as bacteria lack internal membranes whereas eukaryotic cells contain internal membranes such as the endoplasmic reticulum and inside organelles such as mitochondria. List two advantages of having internal membranes.

_____

_____

_____

_____

4 Describe how pH affects the action of enzymes on their substrates.

_____

_____

_____

_____

_____

5 Salivary amylase is an enzyme that breaks down starch into maltose. Its optimum temperature for activity is 37°C

a Give a reason why the optimum temperature for human salivary amylase is 37°C.

_____

_____

b Samples of human salivary amylase and starch were placed together in two test tubes. One of the test tubes was put in a water bath of 70°C. The other test tube was put in a water bath of 10°C. Predict the effect of these temperatures on the amount of maltose produced.

Test tube at 70°C

_____

Test tube at 10°C

_____

c Both test tubes were then put in a water bath of 37°C. Predict the effect on the amount of maltose produced for each test tube. Explain your answer.

_____

_____

d Factors other than temperature affect the rate of enzyme-catalysed reactions. Name two of them.

_____

_____

9780170411660

4 Energy and metabolism

LEARNING

Summary

▶ Adenosine triphosphate (ATP) is the main energy-carrier in a cell. When one of the phosphates of the ATP is removed, adenosine diphosphate (ADP) is produced. Energy is released in this process.

▶ Photosynthesis is a series of chemical reactions in which light energy is used to break down water and carbon dioxide molecules, and build them up into oxygen, glucose and water molecules.

▶ Chloroplasts are the site of photosynthesis.

▶ The photosynthetic reaction is divided into two distinct stages: the light-dependent stage and the light-independent stage.

▶ In the light-dependent stage of photosynthesis, chlorophyll molecules in the thylakoid membrane of a chloroplast absorb light energy. The energy is used to split water molecules into hydrogen ions and oxygen gas.

▶ In the light-independent stage of photosynthesis, hydrogen ions and carbon dioxide are combined to produce glucose in the stroma of a chloroplast.

▶ Aerobic cellular respiration is a series of chemical reactions in which cells break down glucose and oxygen molecules and build up carbon dioxide and water molecules.

▶ In the first stage of aerobic cellular respiration, glycolysis occurs in the cell cytosol. Glucose is partially broken down to pyruvate in enzyme reactions that do not require oxygen, producing two ATP molecules for each glucose molecule.

▶ In the second stage of aerobic cellular respiration, pyruvate enters the mitochondrial matrix where the Krebs cycle occurs. Carbon dioxide molecules are produced. Two ATP molecules are produced for each glucose molecule.

▶ In the third stage of aerobic cellular respiration (electron transfer chain), enzymes in the cristae of the mitochondria facilitate oxygen molecules joining with hydrogen to make water. There is a net yield of 36–38 ATP molecules for each glucose molecule.

▶ Anaerobic cellular respiration is a series of chemical reactions without oxygen, occurring in the cytoplasm.

▶ In anaerobic cellular respiration, pyruvate, produced from glycolysis, undergoes a series of chemical reactions to produce ethanol and carbon dioxide (alcohol fermentation) or lactic acid (lactic acid fermentation). There is a net yield of two ATP molecules for each glucose molecule.

4.1 | ATP–ADP cycle

ATP is an energy-carrier in all living cells. It couples energy-releasing reactions with energy-requiring ones.

1 Draw a diagram of an ATP molecule and an ADP molecule. Use the symbols 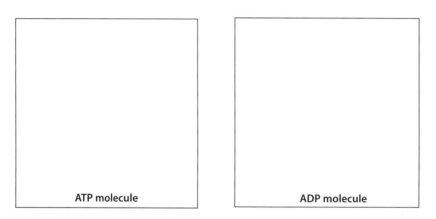 to represent an adenosine molecule and (P) to represent a phosphate molecule.

ATP molecule	**ADP molecule**

2 Complete the equation by writing in appropriate words or circling an alternative.

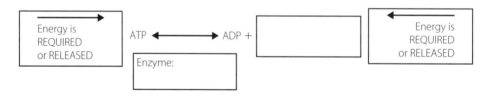

| Energy is REQUIRED or RELEASED | ATP ⟷ ADP + | | Energy is REQUIRED or RELEASED |

Enzyme:

3 Identify which of the molecules ATP or ADP contains more stored energy.

_____

4 State where in the molecule the energy is stored.

_____

5 Explain the process by which energy is used or released in the ATP–ADP cycle. In your explanation, use the knowledge that energy is released when chemical bonds are broken and energy is used when chemical bonds are formed.

_____

_____

_____

6 Explain why the ATP–ADP reaction is referred to as an ATP–ADP cycle.

_____

_____

_____

7 Using the information from this activity, create a diagram of the ATP–ADP cycle. Include the source of the energy for the ATP–ADP cycle and how the energy released is used in a cell.

4.2 | Chemical reactions: photosynthesis

Photosynthesis is a series of chemical reactions catalysed by enzymes that occurs in the chloroplasts of plant cells. It occurs in two different stages: light-dependent and light-independent.

QUESTIONS

1 Using the diagram in Figure 4.2.1, explain what happens in each of the stages.

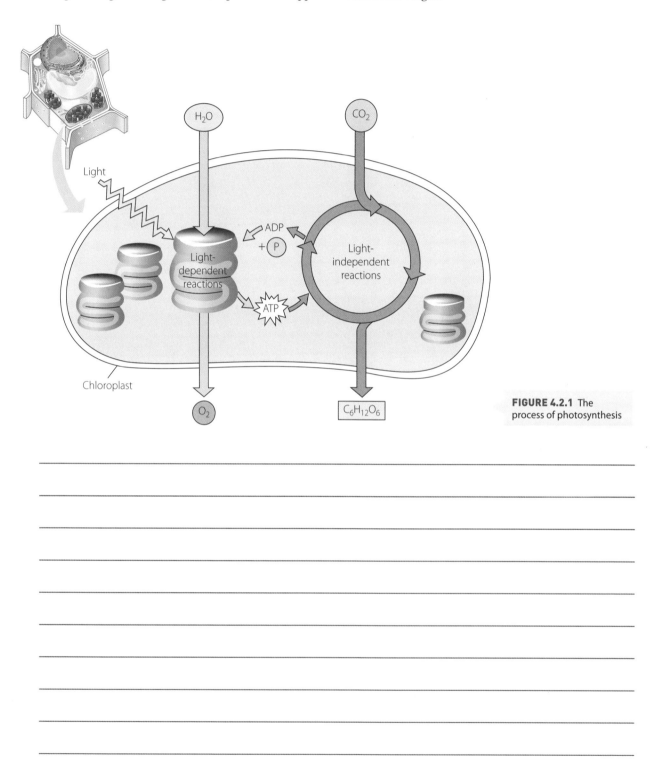

FIGURE 4.2.1 The process of photosynthesis

_____

_____

_____

_____

_____

_____

_____

_____

_____

_____

9780170411660

2 A student wanted to investigate the light-dependent stage of photosynthesis. The diagram in Figure 4.2.2 shows the student's experimental set-up.

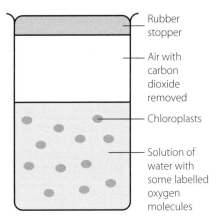

Rubber stopper

Air with carbon dioxide removed

Chloroplasts

Solution of water with some labelled oxygen molecules

FIGURE 4.2.2 Experimental set-up of the light-dependent stage of photosynthesis

The hydrogen acceptor, which was placed in the water solution, is a blue dye that goes colourless when reduced. The beaker was exposed to natural light for 30 minutes. Afterwards, the air space was tested and labelled oxygen molecules were present.

a Explain the presence of labelled oxygen in the air space.

_____

_____

_____

_____

b What is the expected resulting colour of the solution? Explain.

_____

_____

_____

_____

4.3 | Chemical reactions: aerobic cellular respiration

Cellular respiration is the reaction of glucose, oxygen and water to release energy. There are about 20 reactions in this biochemical pathway, all regulated by specific enzymes. Glycolysis is the first pathway that starts the breakdown of glucose. The product from glycolysis, pyruvate, then reacts with oxygen. The by-products are carbon dioxide and water with the formation of ATP.

QUESTIONS

1 Write the word equation for aerobic cellular respiration.

_____

2 Write the balanced chemical equation for aerobic cellular respiration.

_____

3 Complete the table for the inputs and outputs of aerobic cellular respiration.

STAGE OF AEROBIC RESPIRATION	INPUT	OUTPUT	ATP MOLECULES FORMED FROM ONE GLUCOSE MOLECULE	LOCATION IN CELL
Glycolysis				
Kreb's cycle				
Electron chain transfer				

4 Label the diagram in Figure 4.3.1 with words from the word list.

mitochondrion	process of glycolysis	oxygen	cytoplasm
34 ATP	pyruvate	carbon dioxide	glucose
water	2ATP		

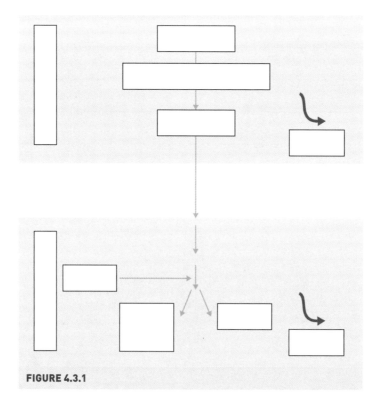

FIGURE 4.3.1

9780170411660

4.4 Chemical reactions: anaerobic cellular respiration

If oxygen is not available, the Kreb's cycle of cellular respiration in the mitochondria does not proceed. Pyruvate, the breakdown product of glycolysis, proceeds down a different pathway of either alcohol fermentation (in bacteria and yeast) or lactic acid fermentation (in animals).

QUESTIONS

1 Write the word equation for alcohol fermentation.

_____

2 Write the word equation for lactic acid fermentation.

_____

3 A man was training for a 15 km fun run. He noticed that after 10 km, his legs started to ache. Explain what is happening inside his muscle cells.

_____

_____

_____

_____

_____

4 The mitochondrion is said to be the site of all cellular respiration. Justify whether you believe this statement is correct.

_____

_____

_____

_____

5 Compare the amounts of energy made available to cells from aerobic respiration and anaerobic respiration.

_____

_____

6 Fill in the boxes in Figure 4.4.1 to compare aerobic and anaerobic cellular respiration.

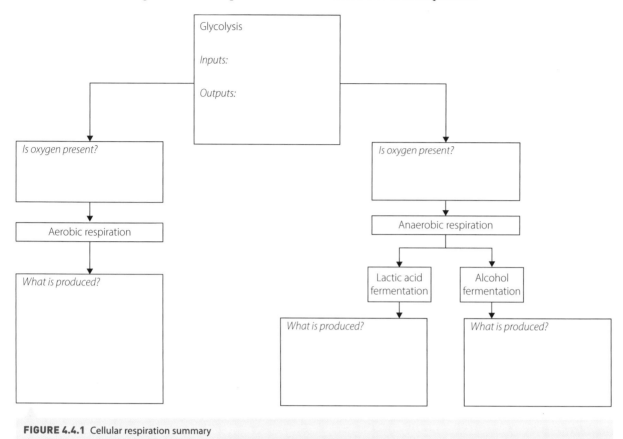

FIGURE 4.4.1 Cellular respiration summary

4.5 | Photosynthesis and cellular respiration

The outputs of photosynthesis are the inputs of aerobic cellular respiration and the outputs of aerobic cellular respiration are the inputs of photosynthesis. The two processes can occur in the same cells in plants.

1 Draw a flowchart showing that the products of photosynthesis and aerobic cellular respiration are used in both pathways.

9780170411660

2   A watered pot plant was placed in an airtight container. The container was illuminated with different intensities of light (lux readings). The amount of oxygen in the container was measured at the different light intensities. A positive O_2 measurement indicates that O_2 is produced (output) and a negative O_2 measurement indicates O_2 is being used (input).

Table 4.5.1 shows the results.

TABLE 4.5.1 O_2 levels at different light intensities

LIGHT INTENSITY (LUX)	O₂ LEVELS (ARBITRARY UNITS)
0	−1
5000	0
10 000	1
15 000	2
20 000	2.5

a   Display data from Table 4.5.1 as a line graph.

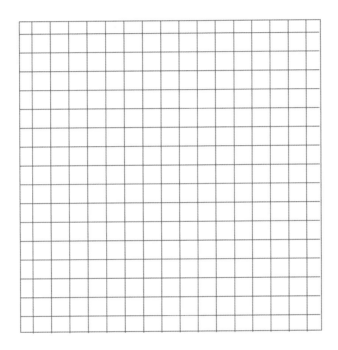

b   Explain how the rates of aerobic cellular respiration and photosynthesis affect the oxygen inputs and outputs.

_____

_____

c   With a coloured pencil/highlighter, shade in the area of the graph where the photosynthetic rate is greater than the aerobic cellular respiration rate. With a different coloured pencil/highlighter, shade in the area of the graph where the photosynthetic rate is less than the aerobic cellular respiration rate.

d   Predict how the rate of photosynthesis compares to the rate of aerobic cellular respiration at the point when oxygen levels are 0.

_____

e   Explain why oxygen exchange is used as a measure of both the rate of photosynthesis and the rate of cellular respiration.

_____

_____

1 Plants use oxygen:

   A during the later stages of respiration

   B during the glycolysis stages of respiration

   C to break down ATP to ADP and inorganic phosphate

   D to combine with carbon dioxide to produce glucose.

2 Name where glycolysis occurs in an animal cell. Compare this with the location of glycolysis in a plant cell.

   _____

   _____

3 Identify the time during a 24-hour day when cellular respiration occurs.

   _____

4 Name the stage of photosynthesis that occurs in the:

   a stroma of chloroplasts _____

   b thylakoid membranes of chloroplasts. _____

5 Compare the products of fermentation in bacteria and yeast with animal cells. Using this knowledge, explain why yeast is used for bread making.

   _____

   _____

   _____

   _____

   _____

6 Explain why ATP is referred to as the main energy-carrier in a cell.

   _____

   _____

   _____

7 You are given two tubes that contain yeast cells. Devise an experiment that would show which tube contains yeast cells that can carry out aerobic cellular respiration only and which tube contains yeast cells that can carry out anaerobic cellular respiration.

_____

_____

_____

_____

_____

_____

_____

8 Some species of algae live in the outer layers of the bodies of jellyfish. The products of algal photosynthesis are absorbed by the jellyfish. If the jellyfish are kept in the dark, predict the most likely outcome for the:

a algae

_____

_____

_____

_____

b jellyfish

_____

_____

_____

_____

5 Cell differentiation and specialisation

LEARNING

Summary

▶ Multicellular organisms consist of a large variety of specialised cells that are structurally different, and which suit the different functions they perform. This specialisation allows the whole organism to function more efficiently, ensuring greater chances of survival and reproduction for individuals, and continuation of the species.

▶ Specialised cells originate from stem cells.

▶ Stem cells differ from other cells because they:
  • have not yet developed into particular types
  • can differentiate to form structurally different specialised cells that perform particular functions
  • have potential to divide and replicate for extended time periods.

▶ Stem cells are named according to their potential to differentiate into different cells.
  • Totipotent stem cells develop into all body cell types of the embryo and tissues supporting it.
  • Pluripotent stem cells develop from totipotent cells into all adult body cell types.
  • Multipotent stem cells develop from pluripotent cells into more than one body cell type within tissues performing a particular function.

▶ The zygote is the first totipotent stem cell of a new individual. The zygote:
  • is formed by the union of sex cells, usually a sperm cell fertilising an egg cell
  • divides to develop into all the different cell types.

▶ Embryonic stem cells are pluripotent cells that develop into the embryo.

▶ Adult stem cells are multipotent cells that develop into different cell types within organs or tissues.

▶ All body cells of multicellular organisms have identical genetic material. Cell differences are determined by which of their genetic instructions are 'switched on'.

▶ Cellular specialisation allows:
  • minimal duplication of tasks in cells
  • a wider variety of biochemical functions
  • more complex physical and cognitive tasks to be performed.

▶ Multicellular specialisation requires communication and coordination between cells. Each cell type is totally dependent on the activities of other cells.

▶ Organ systems are different organs jointly performing a particular function.

▶ Organs are different tissues jointly performing a particular function.

▶ Tissues are specialised cells jointly performing a particular function.

9780170411660

5.1 Properties of stem cells

Unspecialised cells that are able to differentiate into the huge variety of specialised cells are called stem cells. Stem cells also have the potential to divide and replicate for long periods of time. Not all stem cells have the same potential to differentiate, and stem cells are named according to the extent of this ability. The human zygote is the primary stem cell and has total potency; it is totipotent. That is, its continued division produces cells that can develop into any type of cell, including cells that form the placenta to support the developing embryo. Embryonic stem cells develop from the zygote, and can specialise to become all the different cell types in the body; they are pluripotent. Adult stem cells are multipotent. They have more limited potential for specialisation, differentiating into just a certain number of cell types within an area of the body where cell renewal is necessary; for example, bone marrow, skin and the liver.

Complete the following sentences by choosing from the word list.

replicating	totipotent	differentiate	embryo
zygote	adult	unspecialised	

Stem cells are different from other body cells in three ways.

a They are _____ , having not yet developed into a particular cell type.

b They are capable of _____ for a long period of time.

c Stem cells can _____ to form different specialised cells. The _____, formed by the fertilisation of an egg by a sperm, is _____ because it can develop into any type of human cell. Embryonic stem cells are so named because they can develop into any type of cell present in the _____. Embryonic stem cells are pluripotent. _____ stem cells can be called multipotent because they are only capable of differentiating into a limited number of cell types within a particular tissue or organ.

5.2 Stem cell differentiation

Every cell in a developing embryo and in the matured adult organism carries exactly the same set of genetic material, in its nucleus, that was produced by mitotic cell divisions of the original zygote cell. Yet specialised cells, which have different structural features that perform different functions, differentiate in the embryo and persist into adulthood.

At particular stages of development in the embryo, only certain genetic instructions are enabled or 'switched on' to produce each specialised cell type.

The resulting cellular specialisation in a multicellular organism means that individual cells cannot survive without the cooperation and combined functioning of all other cells in the individual.

QUESTIONS

1 What happens within the identical genetic material across different cells of a multicellular organism that enables each cell type to have its own unique specialisations?

_____

_____

_____

_____

2 Choose a particular cell type in the human body and explain how different types of cells in different tissues and organs are needed for your chosen cell type to survive.

_____

_____

_____

_____

5.3 │ Structural hierarchy in multicellular organisms

All living things consist of one or more cells. Cells vary greatly in shape, size, structure and functions performed, and also in how they are organised. Cells are either prokaryotic or eukaryotic. Eukaryotic cells contain membrane-bound compartments, called organelles, which separate areas where particular functions occur. Prokaryotic cells do not have organelles. Prokaryotic cells only occur singly, and include a huge variety of bacteria. Eukaryotic cells can occur singly, such as _Amoeba_, found in fresh water, or as colonies of single cells working together, such as _Volvox_. Eukaryotic cells can also be organised into multicellular organisms composed of groups of very specialised cells arranged in a hierarchy.

Specialised cells that together perform particular functions are called tissues. Different tissues performing particular functions form organs. In complex multicellular organisms, several different body systems are each composed of multiple organs. These systems carry out different roles, and are all totally dependent on each other. For example, in humans, cardiac muscle and nervous tissues occur in the heart organ, which, together with various blood vessel organs, makes up the circulatory system. Essential inputs from the digestive and respiratory systems are transported to every cell in the organism via the circulatory system.

1 Use the words and phrases in Table 5.3.1 to fill the empty boxes in Figure 5.3.1.

TABLE 5.3.1 Words and phrases to fill numbered empty boxes in Figure 5.3.1

Multicellular	Completely functional single cell	Example: circulatory system
Eukaryote	Collection of organs working together to perform particular function/s	Example: bacteria
Organ	Group of specialised cells working together to perform particular function/s	Example: heart
	No organelles	Example: _Volvox_

Use the gradations of white, light grey and dark grey in Table 5.3.1 and Figure 5.3.1 to assist your choice.

9780170411660

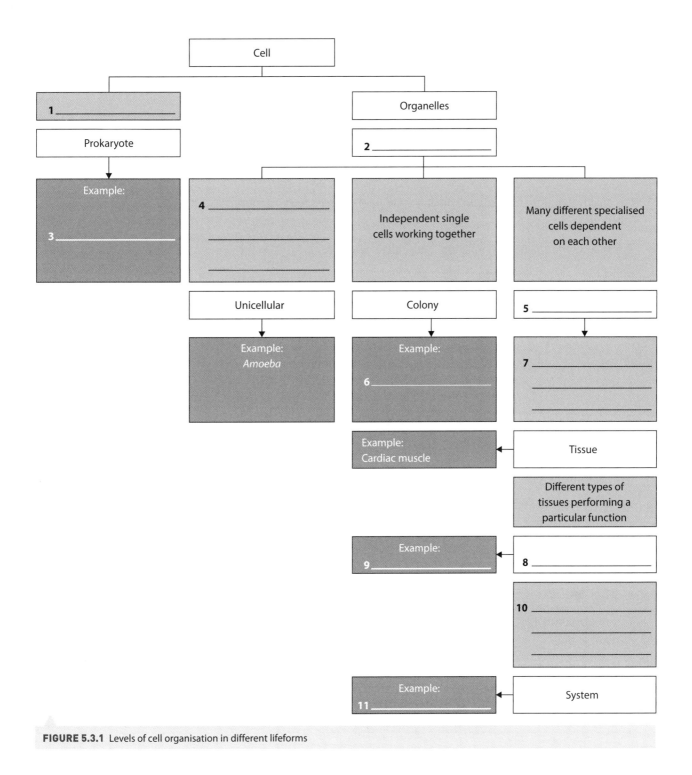

FIGURE 5.3.1 Levels of cell organisation in different lifeforms

2 Use a simple flowchart to illustrate the hierarchical relationship between organisms, cells, systems and tissues in multicellular organisms.

EVALUATION

1 It is an advantage for multicellular organisms to have different types of specialised cells that perform different functions because:

   A it allows the organism to function more efficiently

   B there is less energy wasted overall if all cells don't have to carry out the same functions

   C it allows a greater range of functions, including complex ones, to be carried out overall among all the cells

   D all of the above.

2 In multicellular organisms, cells are arranged in a hierarchy where:

   A tissues are grouped into cell types

   B groups of cells performing a particular function are called organ systems

   C groups of cells performing a particular function are called tissues

   D many organs form tissues.

3 All the different specialised cells in a multicellular organism originate from _____ cells.

4 Stem cells are different from all other cells because they:

   • have the ability to continue to _____ over very long time periods

   • have not yet _____ into a particular specialised cell type

   • are capable of developing into cells with different structures to perform different _____ when they do differentiate

5 Every cell of a multicellular organism has exactly the same genetic material.

   Briefly explain how it is possible then for different cells to have different structural features.

   _____

   _____

   _____

9780170411660

6 Gas exchange and transport

LEARNING

Summary

- Cells of all organisms require inputs for, and produce outputs from, all the chemical reactions they perform. Some of these inputs and outputs are gases.

- In animals, and in plants not photosynthesising, oxygen gas (O_2) is a necessary input for cellular respiration, the series of reactions that produce usable energy. Carbon dioxide gas (CO_2) is a product that must be eliminated.

- Continual gaseous exchange occurs with the external environment because gases cannot be stored.

- Gas exchange with the external environment occurs across specialised surfaces that must:
  - be moist for CO_2 and O_2 to dissolve in water (H_2O) to diffuse across the surface
  - be thin and permeable
  - have a large surface area to exchange adequate gaseous quantities
  - have a greater concentration of gas on one side than the other for diffusion to occur.

- In mammals, the surface for gas exchange with the outer environment is the alveolus wall, deep inside the lung. The alveolus is part of the whole respiratory system; its internal location is necessary to stop the surface dehydrating.

- In fish, the exchange surface is the gill membrane, on the outside of the body.

- The exchange surface must also be extremely close to the transport system (capillaries of circulatory system) for delivery of gases to all body cells.

- The capillary wall is a specialised surface through which gases pass during exchanges between the internal and external environments.

- The capillary wall has the four specialised characteristics for gas exchange, as listed above.

- The capillary has a tubular shape with a smooth inner surface and is ideally structured also for transportation of blood, which is the liquid in which gases dissolve.
  - O_2 attaches to haemoglobin molecules in red blood cells.
  - CO_2 mostly dissolves in water in blood, but some is carried by haemoglobin.

- The extremely small diameter of capillaries and their extensive network allow gaseous delivery to, and pick-up from, every cell in body.

- The heart is a muscular pump that propels blood through the variously sized vessels of the closed circulatory system.

6.1 | Structure and function of gaseous surfaces

Figure 6.1.1 illustrates the two single flat cell layers, and intercellular fluid between them, that O_2 molecules diffuse across, from the inhaled air of the external environment in the alveolus, into the internal body environment of the blood inside the capillary. At the same time, CO_2 diffuses in the opposite direction.

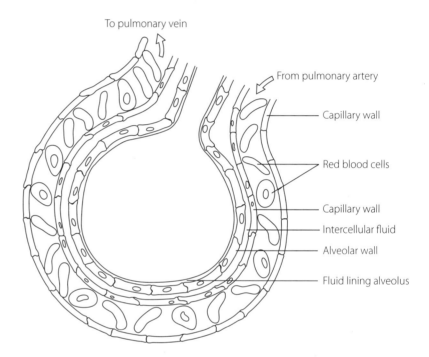

FIGURE 6.1.1 Structural detail of an alveolus and a capillary, at cellular level

Refer to Figure 6.1.1 to complete the following questions.

Conventionally, *red* indicates more highly oxygenated blood, and *blue* indicates deoxygenated blood; blood with a lower concentration of O_2.

1 Colour the 'From pulmonary artery' and 'To pulmonary vein' arrows appropriately.

2 Briefly describe how the pulmonary artery is different from any other artery.

_____

_____

3 Appropriately colour the four red blood cells closest to each of the arrows in Figure 6.1.1.

4 Add your own arrows to illustrate the directions in which you would expect O_2 and CO_2 to diffuse across the walls of the alveolus and capillary.

5 Complete the following flowchart showing the structures through which inhaled air passes between the mouth/nose and the alveolus:

Mouth/nose → throat (pharynx) → voice box (larynx) → t_____ → bronchus → b_____
→ alveolus

6 Draw a simple labelled diagram of the human respiratory system, showing right and left lungs, and relative positions of each of the last four structures in the above list.

7 Briefly describe the structural *similarities* between alveoli in the lungs of mammals and gills in fish that allow them both to function as gaseous exchange sites.

_____

_____

_____

_____

8 Briefly describe the *difference/s* between alveoli and gills.

_____

_____

_____

_____

9 Briefly explain how each difference suits each organism to its particular environment.

_____

_____

_____

_____

6.2 Structure and function of capillaries

Capillaries have three major functions.

▶ Capillaries surrounding alveoli have walls that are exchange surfaces for O_2 and CO_2 gases between air of external environment in alveoli and blood of internal environment; circulatory and respiratory systems function jointly to acquire necessary O_2 and eliminate waste CO_2 for all body cells.

▶ All other capillaries have walls that are exchange surfaces to deliver required materials *to* and remove waste materials *from* body cells.

▶ Capillaries are components of an extensive, closed transportation 'pipeline' linking all body cells, tissues and organs, and:

- are the finest 'pipes' that branch from larger vessels that carry blood pumped from the heart
- have multiple branches that form intricate networks inside organs and tissues throughout the entire body that are within microscopic distance of every body cell
- these networks then progressively converge into larger vessels returning blood to the heart.

Larger vessels include arteries and smaller diameter arterioles, which carry blood *away from* the heart, and veins and finer venules, which transport blood *to* the heart.

1   Produce a three-dimensional diagram of part of a capillary that illustrates the jigsaw-like nature of epithelial cells that all fit together to form a thin, smooth 'pipe'.
    Draw:

▶ a circle for the opening of the capillary

▶ two parallel lines from opposite sides of the circle in an upwards left direction

▶ thin flattened cross-sections of three cells meeting around the margin of the circle, ensuring two of these meet between the two lines extending off the circle

▶ a slightly raised 'lump' in each cross-sectional cell

▶ a cross-sectional view of nucleus within each lump

▶ a flattened surface view of cells to complete the jigsaw puzzle-like surface pattern of the capillary between the two lines.

9780170411660

2 Complete the table for the structure and functions of capillaries.

CAPILLARY STRUCTURE	FUNCTION
Smooth inner surface	
	Easily allows diffusion
Extensive branching	
Very small diameter	

1 One essential characteristic of gas exchange surfaces – alveoli in mammals, and gills in bony fish – is large surface area. This is achieved by:

A very thin, single-cell thickness membranes in both

B close proximity to capillaries in both

C multiple layers of feathery filaments in alveoli, multiple bunches of air sacs in gills

D multiple bunches of air sacs in alveoli, multiple layers of feathery filaments in gills.

2 The reason why oxygen gas is continually taken in from the external environment, but food intake is not a continuous process is that:

A gases cannot be stored within the body but food breakdown products can

B food breakdown products cannot be stored within the body but gases can

C oxygen gas is more essential to cells than food breakdown products

D oxygen gas is needed in much larger quantities relative to the quantities of food required.

3 Water flows over the gills of a bony fish in the opposite direction to the flow of blood in the capillaries inside. This process increases gaseous exchange quantities and is called 'c_____ c_____ exchange'.

4 The complex protein inside red blood cells that bonds with oxygen molecules to transport them to cells is called h_____.

5 The capillary wall is the exchange surface through which many molecules pass between the blood and the body cells by diffusion.

a List three of these molecules.

_____

b Define:

i diffusion

_____

ii concentration gradient

_____

9780170411660

LEARNING

Summary

▶ In addition to gaseous oxygen, the other essential input for cellular respiration reactions to produce energy in the form of adenosine triphosphate (ATP), is an organic 'food' source, in the form of other living things. This food source also supplies all necessary inputs for the multitude of other chemical reactions occurring in all body cells.

▶ Non-useful ingested materials and metabolic wastes must also be eliminated.

▶ Food from the external environment is taken into the digestive system but cannot pass into the internal environment until it is digested to molecules small enough to pass across the exchange surface into the circulatory system for distribution to all body cells.

▶ Chemical (and some mechanical) breakdown occurs progressively along the digestive tract.
  • Protease enzymes break down proteins to amino acids.
  • Lipases digest lipids to fatty acids and glycerol.
  • Carbohydrases, including amylases and maltase, digest carbohydrates to monosaccharides.

▶ Digestion products diffuse across the one-cell thick wall of villi, lining the small intestine, then across a very small distance of intercellular fluid.
  • Amino acids and saccharides (simple sugars) then diffuse across the one-cell-thick capillary walls and into the circulatory system.
  • Digested lipids diffuse across the exchange surface of lacteals instead of capillaries and into the lymphatic system, which later empties into the circulatory system.

▶ Undigested materials continue along the digestive tract and are eliminated as faeces.

▶ Excess carbohydrate and lipid breakdown products are stored for later use, but excess amino acids are broken down into usable carbon-containing molecules and (non-useful) nitrogenous waste molecules.

▶ Nitrogenous wastes are converted to urea in the liver and eliminated from blood, along with other waste and excess molecules, as urine.

▶ Urine is produced in nephrons, the working units of the kidney.

▶ Filtration, selective reabsorption and secretion are carried out by various specialised nephron structures, including the glomerulus, Bowman's capsule, proximal tubule, loop of Henle, distal tubule and collecting tubule.

7.1 Absorption of nutrients

The inner lining of the small intestine is folded and ridged, and its surface area is increased further by being completely covered with minute, elongated projections called villi (singular: villus). (Latin *villi* = 'shaggy hair'). The walls of the villi are composed of a single layer of epithelial cells, and the exposed membrane of these cells also has a convoluted surface of microvilli. Protruding up inside each villus is a branching capillary network, and also a lacteal vessel. The exiting capillaries converge into venules, and the lacteals into lymph vessels, which then carry the products of digestion to the liver, where they are processed in various ways. Further digestion of fats, conversion and storage of excess glucose, and detoxification of harmful ingested materials and metabolic products, are some of the liver's functions.

1   Draw and label a diagram of a villus, illustrating its wall made up of a single layer of columnar epithelial cells and the capillary network and lacteal vessel inside it.

2   Add the following details to the diagram from Question 1.

   a   Draw in the arteriole from which the entering capillary branches, the venule into which the exiting capillary leads, and the lymph vessel from which the lacteal extends.

   b   Colour the arteriole and venule appropriately (red or blue, see Chapter 6), and also appropriately colour the section of the capillary network carrying oxygenated blood into, and the section of the network carrying deoxygenated blood out of the villus.

   c   Add arrows to show entering and exiting capillaries.

9780170411660

7.2 | Digestion of food

Chemical digestion of the different types of organic matter happens in stages, and within different areas of the digestive tract. Digestion is mainly complete by the time the material reaches the jejunum (second/middle section) of the small intestine. Most of the absorption of nutrients occurs from the jejunum and the ileum (last section) of the small intestine.

Complete the table showing the organs of digestion.

ORGAN		IS ORGAN A SITE OF CHEMICAL DIGESTION? (YES OR NO)	IF YES	
			TYPE OF ORGANIC MOLECULE	ENZYME TYPE
MOUTH				
OESOPHAGUS				
STOMACH				
SMALL INTESTINE	DUODENUM			
	JEJUNUM			
	ILEUM			

7.3 Nitrogenous wastes

The liver processes excess sugars and lipids for storage so that they can be used for energy when not enough energy containing food is ingested. However, in the metabolism of excess amino acids, the amino parts of the molecules containing nitrogen are removed for excretion, leaving the non-nitrogenous parts to be converted into sugars and lipids.

Ammonia, uric acid and urea are the three different products of nitrogen waste breakdown that organisms excrete, depending on which suits their particular way of life.

1   Write the letters of the phrases that apply to each nitrogenous waste. (Phrases may apply to more than one nitrogenous waste.)

Nitrogenous wastes	Letters of phrases that apply
Ammonia	_____
Urea	_____
Uric acid	_____

Phrases:

A Nitrogenous waste excreted by reptiles, birds and insects

B Nitrogenous waste excreted with maximum water loss

C Extremely toxic nitrogenous waste

D Nitrogenous waste excreted as paste with minimal water loss

E Nitrogenous waste excreted by humans

F Nitrogenous waste excreted by fish

G Less toxic product of ammonia

9780170411660

7.4 Nephrons

Waste from every cell of the entire body is transported in the circulatory system, and the main organs that function to excrete this waste are the kidneys. Approximately one-quarter of all blood leaving the heart is sent to the kidneys in the renal artery, where about 1.2 litres of blood fluid is filtered every minute.

Nephrons are the specialised functional units of the kidney that carry out this complex task, and also ensure water, solute and acidity levels of the body fluids remain within optimal levels for overall body functioning.

Draw and label a diagram of the nephron, including the following structures.

| glomerulus | Bowman's capsule | proximal tubule |
| loop of Henle | distal tubule | collecting tubules |

7.5 | Production of urine

Filtration occurs at the glomerulus, where about one-fifth of the blood plasma (water) and all the substances within the plasma, except blood cells and large proteins, are forced out by high pressure through the capillary wall, then through the Bowman's capsule wall and into the proximal tubule.

Selective reabsorption occurs when:

▶ some water, all glucose, amino acids, and around 65% of mineral ions pass back out of the proximal tubule into the surrounding capillaries

▶ more water passes passively out of the descending loop of Henle due to active (and later passive) transport of minerals out of the ascending section of loop of Henle, increasing mineral ion concentration outside tubules in intercellular fluid and capillaries

▶ more water, if necessary, and more useful minerals move out of the distal tubule

▶ more water is reabsorbed from the collecting tubule if necessary.

Secretion is the process in which any additional nitrogenous waste and any materials in excess of body requirements are transferred out of the capillaries and intercellular fluid and into the nephron tubules already carrying urea and other wastes.

Secretion occurs into the collecting tubules (and perhaps also into the proximal and distal tubules) if necessary.

At all appropriate places on Figure 7.5.1, add your own arrows, with heads showing the direction of movement of materials, and labels as suggested below, to illustrate where the following processes take place.

▶ Filtration (label F)

▶ Reabsorption of water (label Rw), glucose (label Rg), amino acids (label Ra), minerals (label Rm)

▶ Secretion (label S)

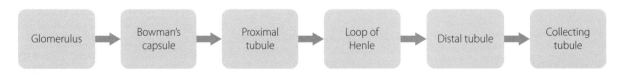

FIGURE 7.5.1 Flowchart showing structures through which waste materials pass in nephron

9780170411660

1 All living things require inputs from the external environment, which need to pass across an exchange surface to reach the body's internal environment.

Choose the correct statement with respect to this idea.

A Exchange surfaces of organisms need to be outside the body.

B Exchange surfaces of organisms can be deep within body.

C All organisms need to protect exchange surfaces from drying out.

D In different organisms, exchange surfaces for gas and for nutrients are totally different in structure.

2 Choose the correct list of organic compounds and their breakdown products.

A Proteins to amino acids; carbohydrates to fatty acids and glycerol; lipids to simple sugars

B Proteins to fatty acids and glycerol; carbohydrates to simple sugars; lipids to amino acids

C Proteins to amino acids; carbohydrates to simple sugars; lipids to fatty acids and glycerol

D Proteins to amino acids and fatty acids; carbohydrates to simple sugars; lipids to glycerol

3 The alveoli, villi and nephrons are sites where exchange of materials between external and internal environments of an organism take place.

For each of the alveolus, villus and nephron, discuss their particular structures and functions by including:

a which parts of them enclose environment external to and which parts contain environment internal to the body

b the actual surfaces across which exchanged molecules move between external and internal environments.

Also discuss:

c similarities between the three structures that suit them to their general role for molecular exchange

d differences between the three structures that suit each to their specific function.

_____

_____

_____

_____

_____

_____

_____

_____

_____

_____

8 Plant systems: gas exchange and transport systems

LEARNING

Summary

▶ The stomata and guard cells control the movement of gases in leaves by opening when they are turgid and closing when they are flaccid.

▶ Air spaces between the cells in the spongy mesophyll allow gases to enter through the stomata and diffuse throughout the leaf without energy expenditure.

▶ The palisade mesophyll contains the bulk of the chloroplasts and are tightly packed on the upper surface of the leaf, which maximises light exposure. The spongy mesophyll contains the rest of the chloroplasts and are loosely packed on the underside of the leaf, which allows gas exchange and catches reflected light. The upper and lower epidermis and cuticle protect the leaf from excess water loss.

▶ The vascular tissues of leaves include the xylem, which carries water and dissolved minerals from the roots to the leaves, and the phloem, which carries sap and sugars from the leaves to the rest of the living plant tissues. Xylem is constructed from dead cells that have been organised into tracheids and vessel elements, while phloem is constructed from living cells (sieve tubes) that are connected by porous walls called sieve plates. The sieve tubes do not contain a nucleus or most of the organelles and therefore rely on companion cells to survive.

▶ Water uptake from the soil produces pressure within the root xylem, which forces water into the stem. Capillary action caused by both adhesion and cohesion of water molecules lifts water up the stem and the transpiration stream encourages the continual flow of water out through the stomata in the leaves.

▶ The rate of transpiration is raised by increased light intensity, temperature and wind and lowered by increased humidity.

▶ The sugars produced by photosynthesis are transported by translocation in the phloem, which always flows towards areas of high sugar usage such as areas of growth and reproduction.

8.1 | Stomata

Stomata are important structures for a plant's access to gases. Their role in regulating the entrance of fresh air allows plants to balance their gas needs with their water loss. This activity aims to model guard cell function, to give a clearer understanding of how stomata are opened and closed.

Blow up a long skinny balloon; do not tie the end. Bend the balloon in a circle to imitate an opening. This is the turgid position. Slowly let the air out of the balloon while continuing to hold it in the circle. Watch the opening close up. This is the flaccid position.

1   Sketch the stomata in the turgid position. Include descriptive labels to show the cellular conditions that allow the stomata to be open.

2   Sketch the stomata in the flaccid position. Include descriptive labels to show the cellular conditions that allow the stomata to be open.

3 Describe the role of stomata and guard cells in controlling the movement of gases in leaves.

_____

_____

_____

_____

_____

4 Summarise the strengths and weaknesses of this activity in accurately modelling the function of guard cells.

STRENGTHS	WEAKNESSES

9780170411660

8.2 Facilitating gas exchange

Annotate Figure 8.2.1, a diagram of a leaf cross-section.

1   Label the cuticle, epidermis, palisade mesophyll, spongy mesophyll, stomata, guard cells, xylem and phloem.

2   Use red arrows to show the flow of CO_2 within the plant structures.

3   Use green arrows to show the flow of O_2 within the plant structures.

4   Use blue arrows to show the flow of water vapour within the plant structures.

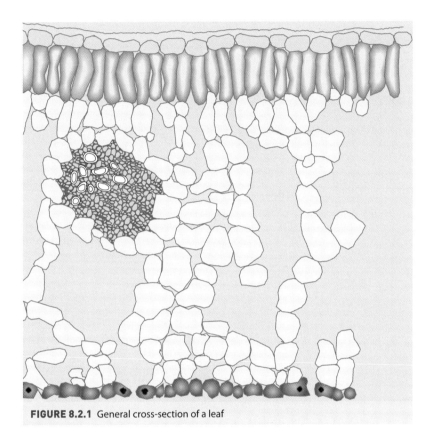

FIGURE 8.2.1 General cross-section of a leaf

8.3 | Relationship between photosynthesis and the main tissues of leaves

Leaves have evolved to be extremely streamlined and efficient photosynthesisers. Every leaf part has an important function in the process. The arrangement of the parts within the larger structure is also very important.

QUESTIONS

1 Complete the table to detail the structures and functions of each component of the leaf.

COMPONENT	STRUCTURE	FUNCTION
Cuticle		
Epidermis		
Palisade mesophyll		
Spongy mesophyll		
Phloem		
Xylem		
Stomata		

9780170411660

2 Explain why the spongy and palisade mesophyll may have evolved as different tissues within the leaf.

_____

_____

_____

_____

3 Use a textbook or other source of accurate information to research how the cross-section of a gum leaf is different from the cross-section of a general leaf. Explain why you think these differences have evolved in gum leaves and how they relate to photosynthesis requirements.

_____

_____

_____

_____

_____

8.4 Xylem and phloem

Complete the table to compare and contrast the two plant vascular systems: xylem and phloem. Use diagrams and colour to clarify and condense your information. Ensure you acknowledge the sources of any diagrams you copy.

FEATURE	XYLEM	PHLOEM
STRUCTURE Describe the cells it is made of and how the cells are arranged together.		
FUNCTION Describe what it transports, in which direction and the forces that drive the movement.		
LOCATION Describe its location within the cross-section of a plant stem. (Include both monocot and dicot stems.)		

8.5 Rate of transpiration

Create a mind map to show the connections between the words in the list. Include relevant information about each of the words and illustrations where appropriate.

Transpiration	Light	Temperature	Wind
Humidity	Rate	Potometer	Water
Evaporation	Stomata		

9780170411660

8.6 Terraria

Terraria (Latin *terra* = 'land') are self-sustaining closed ecosystems, usually sealed into glass containers. A terrarium (Figure 8.6.1) contains soil, water and plant life, and on a sunny windowsill they can grow well for years. Open terraria require the input of water occasionally, but closed terraria need only sunlight.

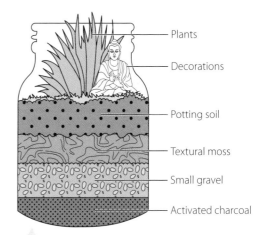

FIGURE 8.6.1 A common terrarium set-up

QUESTION

Discuss how water and gases would be cycled in a closed terrarium, so that the plants could live for years sealed in a small container.

_____

_____

_____

_____

_____

_____

_____

_____

_____

_____

_____

_____

_____

_____

_____

_____

_____

_____

_____

_____

_____

_____

_____

1 Which structure in the leaf facilitates gas entry and exit?

   A Chlorophyll

   B Mesophyll

   C Stomata

   D Vascular bundle

2 Which of the following would increase the rate of transpiration?

   A High temperature

   B Low wind

   C High humidity

   D Low light

3 Name the two vascular tissues of a plant.

_____

4 On the diagram of a tree in Figure 8.7.1, use a blue pen to draw arrows showing the direction of water movement.
   Use a red pen to draw arrows showing the direction of the movement of sugars.

FIGURE 8.7.1 Movement of substances in a plant

5 Explain how a leaf is ideally structured to optimise photosynthesis.

_____

_____

_____

_____

_____

_____

6 A student set up an experiment as shown in Figure 8.7.2. It was left out on an open oval from 7 am to 5 pm on a sunny day. Explain the changes in the concentration of oxygen, carbon dioxide and water vapour in the jars over this period of time.

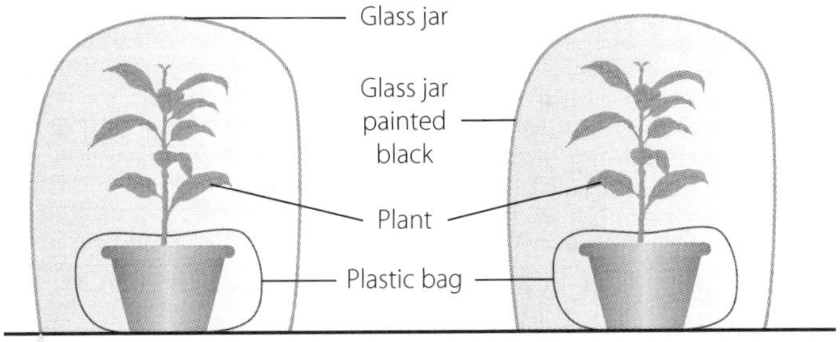

Glass jar

Glass jar painted black

Plant

Plastic bag

FIGURE 8.7.2 Plant transpiration experiment set-up.

_____

_____

_____

_____

_____

_____

_____

_____

_____

_____

_____

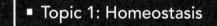

UNIT TWO

MAINTAINING THE INTERNAL ENVIRONMENT

- Topic 1: Homeostasis

- Topic 2: Infectious diseases

9780170411660

9 Homeostasis

Summary

▶ Homeostasis can be described by a stimulus–response model, in which changes in the internal or external environment are detected by receptors, processed by the central nervous system and appropriately responded to by effectors.

▶ The five categories of sensory receptor are chemoreceptors (chemical detection), thermoreceptors (heat detection), mechanoreceptors (pressure detection), photoreceptors (light detection) and nociceptors (pain detection).

▶ Effectors respond to stimuli and either contract, as in muscles, or secrete substances, as in glands.

▶ Homeostasis uses negative feedback to maintain body conditions within a narrow range of acceptable values, referred to as tolerance limits. Negative feedback is characterised by the response inhibiting or counteracting the stimulus, providing a dampening effect on the original stimulus.

▶ Feedback control diagrams depict the chains of stimuli and responses that provide feedback to the body on the results of an action. Both nervous and hormonal pathways are used.

▶ Metabolism describes all of the chemical reactions involved in sustaining life. Catabolism is all of the reactions that produce energy by breaking down large molecules, while anabolism is all of the reactions that use energy to produce large molecules.

▶ Enzymes power almost all metabolic reactions and require specific conditions to work at optimal efficiency. Changes in metabolic activity affect body temperature, blood pH and other factors that alter the efficiency of the enzymes in the body, and so they must be regulated through negative feedback to maintain homeostasis.

9.1 | Stimulus–response model

The stimulus–response model is often linked to homeostasis, but is also vital for responses that are not linked to maintaining life.

Sketch the models for each of the following situations. Include detail where appropriate.

1 Stimulus: increased internal body temperature

2 Stimulus: decreased environmental light

3 Stimulus: something flying at your head

4 Stimulus: increased blood glucose level

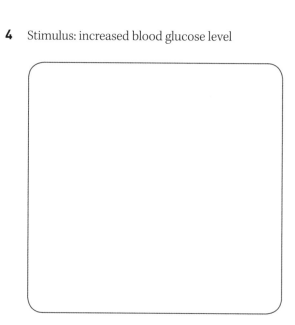

9.2 Sensory receptors

Organisms, and their cells, have many different shaped and sized receptors in order to receive different kinds of information. There are millions of external and internal receptors that allow organisms to respond to stimuli, but there are only five main categories: chemoreceptors, thermoreceptors, mechanoreceptors, photoreceptors and nociceptors.

Complete the table to summarise the general location and function of each of the five categories of sensory receptor.

CATEGORY	LOCATION	FUNCTION
Chemoreceptor		
Thermoreceptor		
Mechanoreceptor		
Photoreceptor		
Nociceptor		

9.3 Effectors

Effectors are the parts of the body that respond to stimuli. They are either muscles or glands. Muscles respond to neural signals by contracting and glands respond to both neural and chemical signals by secreting various substances.

1   Draw a diagram to show the structure of a muscle fibre as it is contracting.

2 Explain the role of the pancreas as an effector in the body.

_____

_____

_____

_____

3 List the substances secreted by the pituitary gland and their effects in the body.

_____

_____

_____

_____

_____

4 Considering your response to Question **3**, explain why the pituitary gland is often called the 'master gland'.

_____

_____

_____

_____

_____

_____

_____

_____

9780170411660

9.4 Feedback control diagrams

Feedback control diagrams are a systematic way to notate the complicated connections of biological responses. Arrows are used to show the direction of the process, while plus and minus signs are used to show whether factors are increased or decreased, respectively. Figure 9.4.1 shows an example of a negative feedback control diagram of the body's response to haemorrhage (excessive blood loss).

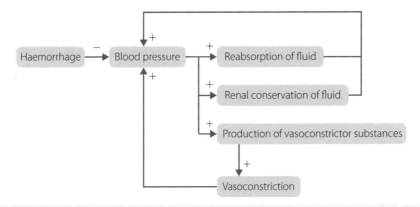

FIGURE 9.4.1 Feedback control diagram showing the body's responses to haemorrhage

1   Study Figure 9.4.1. Summarise which body systems are involved in counteracting haemorrhage and the role they play in this process.

_____

_____

_____

_____

2   Use the negative feedback model to draw a feedback control diagram that explains how the body maintains a constant internal temperature of approximately 37°C. Begin with the stimulus of metabolism, which will increase internal temperature. Ensure you include all systems involved in the process.

9.5 Metabolism

'Metabolism' is the term used to refer to all of the chemical reactions required for life. However, in science, it is mostly used in the adjectival form, as in metabolic reactions and metabolic disorders. In general society, the term 'metabolism' has been skewed to mean something slightly different, and has two forms 'fast' and 'slow'. Unfortunately, these terms can be misleading.

QUESTIONS

1   Define:

   a   metabolism

   _____

   b   catabolism

   _____

   c   anabolism

   _____

   d   basal metabolic rate

   _____

   e   kilojoules/calories

   _____

2   'Fast metabolism' is generally used to refer to people who can eat freely without excess weight gain. Explain why this term is inaccurate.

   _____

   _____

   _____

   _____

   _____

3   The enzymes that drive metabolic reactions require specific conditions to operate. Suggest three ways to improve these conditions in your body to optimise your enzyme activity.

   _____

   _____

   _____

   _____

   _____

9780170411660

1 How many categories of receptor are there?

   A Three

   B Four

   C Five

   D Six

2 Which of the following factors *do not* affect enzyme activity?

   A pH

   B Temperature

   C ATP availability

   D Blood glucose level

3 Give one example of catabolism and one example of anabolism.

   _____

4 Draw a stimulus–response model for the body's response to brushing against a hot stove. Include as much detail as possible.

5 Explain the role of receptors in homeostasis.

   _____

   _____

   _____

6 Define 'positive feedback' and explain why it disrupts homeostasis.

   _____

   _____

   _____

   _____

   _____

10 Neural homeostatic control pathways

LEARNING

Summary

▶ Nerve impulses are transported by three categories of neurons: sensory receptor cells (sensory neurons), interconnecting neurons (interneurons) and effector neuron cells (motor neurons).

▶ Sensory neurons relay signals specifically from the stimulus to the central nervous system (CNS), interneurons relay signals between sensory neurons and motor neurons and motor neurons relay signals between the CNS and the effectors (either glands or muscles).

▶ General neuron structure includes dendrites at one end, a long axon wrapped in fatty myelin, and axon terminals at the other end.

▶ Signals pass from dendrites to axon terminals. The cell body (soma) is located in different parts of the neurons, as shown in Figure 10.0.1.

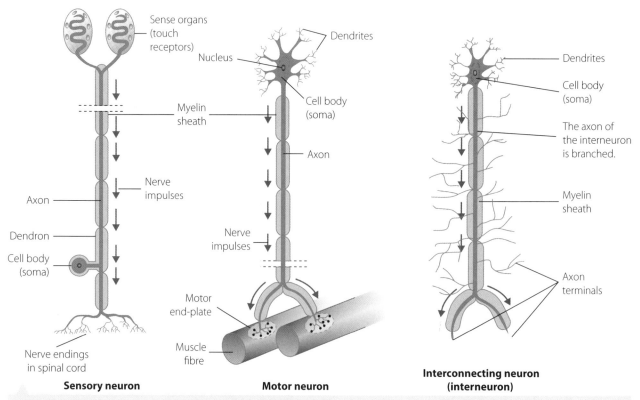

FIGURE 10.0.1 The structures of the sensory, interconnecting and motor neurons

9780170411660

- Schwann cells wrap themselves tightly around the elongated axon to form the myelin sheath. The gaps between Schwann cells are called nodes of Ranvier. The myelin sheath insulates the neuron from other signals and improves transmission speed because the signal must jump between nodes.
- Nerve impulses travel through the neuron in a process called transmission (Figure 10.0.2).

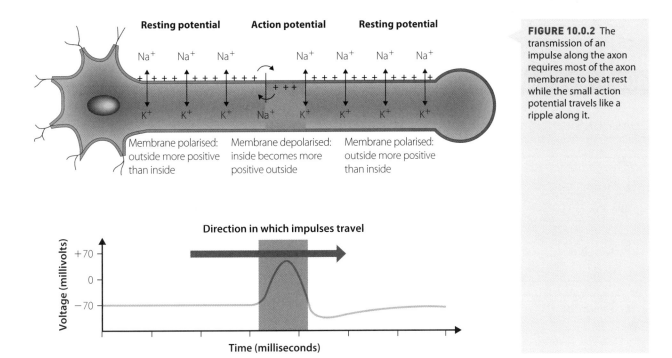

FIGURE 10.0.2 The transmission of an impulse along the axon requires most of the axon membrane to be at rest while the small action potential travels like a ripple along it.

- Nerve impulses travel between neurons by signal transduction (Figure 10.0.3).

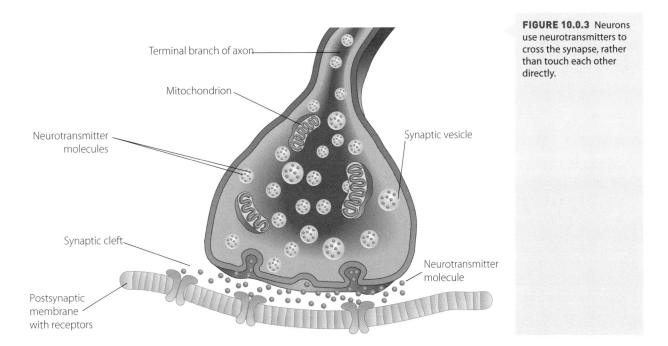

FIGURE 10.0.3 Neurons use neurotransmitters to cross the synapse, rather than touch each other directly.

10.1 | Cells that transport nerve impulses

QUESTIONS

1 Define:

a axon

_____

b soma

_____

c dendrites

_____

d myelin

_____

2 Explain how the myelin sheath is formed in neurons.

_____

_____

_____

3 Neurons inside the brain are wired together like a mesh web, while neurons outside the brain are bundled together in nerve fibres. Often, certain sitting or lying positions can create what is colloquially known as a 'dead leg' or 'dead arm'.

   Use your knowledge of nerve structure and function to explain why your entire leg can go numb from sitting in certain positions too long, and why 'pins and needles' occurs after the pressure is relieved.

_____

_____

_____

_____

_____

_____

_____

9780170411660

10.2 | Comparing three types of neurons

There are three types of neurons: sensory neurons, motor neurons and interneurons. Sensory neurons respond to external stimuli and send signals to the central nervous system. Motor neurons enable involuntary and voluntary muscle functions. Interneurons are sometimes described as the 'middle man' between the sensory and motor neurons.

Use pipecleaners to construct models of each of the three types of neurons. Try to make your models as accurate as possible, include as many features as you can.

1 Contrast a real neuron with your models. How are they similar and how are they different?

SIMILARITIES	DIFFERENCES

2 Compare and contrast the structure and function of your sensory and motor neurons by completing the Venn diagram in Figure 10.2.1.

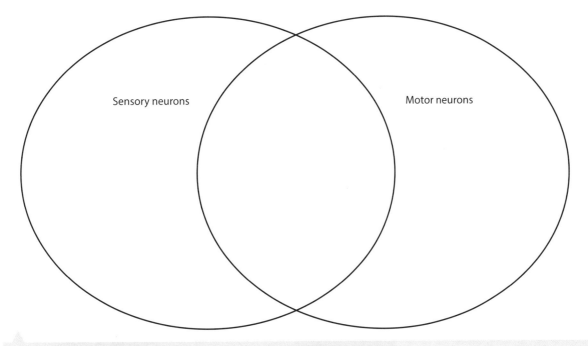

FIGURE 10.2.1 A Venn diagram comparing and contrasting sensory and motor neurons

10.3 | Passage of a nerve impulse

Nerve impulses require two distinct processes to travel through the nervous system – transmission and transduction. Transmission refers to the passage of the electrical impulse from one end of the neuron to the other, while transduction refers to the way that the signal is passed from one neuron to another using chemicals called neurotransmitters. Using a pack of M&Ms or other individual counters, a few sheets of paper and a camera, you can construct a working model of transmission and transduction.

1   Set up the paper along a work surface in two areas (Figure 10.3.1). The paper's edge will delineate the neuron cell membrane. You should have two 'neurons' separated by a 'synapse'.

FIGURE 10.3.1 The 'neurons' separated by their 'synapse'.

2   Select a colour (or other identifying feature of your counters) for each of the following: potassium ions, sodium ions, calcium ions, neurotransmitters and receptors.

3   Photograph each of the following stages from as close to the same angle as possible.

4   Begin with both neurons in resting potential (three sodium ions outside the cell for every two potassium ions inside the cell) (Figure 10.3.2).

FIGURE 10.3.2 Neurons at rest have more sodium ions outside the cell than potassium ions inside.

9780170411660

5 Trigger an impulse in the left-most neuron, which will result in an influx of sodium ions to create an action potential (Figure 10.3.3).

FIGURE 10.3.3 Action potential is when there are more positive ions inside the cell than outside.

6 Move the action potential down the neuron by swapping sodium for potassium ions. Make sure you reset the neuron behind your action potential by pushing the sodiums back out again (Figure 10.3.4).

FIGURE 10.3.4 The nerve impulse flows down the axon.

7 Upon reaching the end of the neuron, take some time to set up the synapse and right-most (receiving) neuron for transduction. The left-most neuron will have small vesicles of neurotransmitters, the synapse will have sodium and calcium ions and the receiving neuron will have receptors on its cell membrane (Figure 10.3.5).

FIGURE 10.3.5 The synapse at rest

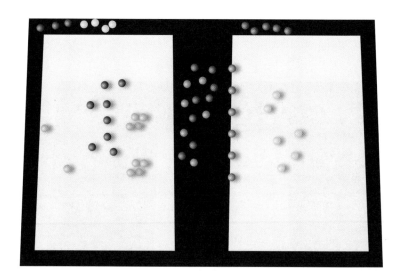

8 The sodium ions filling the left-most neuron open calcium ion channels in the membrane. Move the calcium ions into the left-most neuron to trigger exocytosis of the neurotransmitter vesicles (Figure 10.3.6).

FIGURE 10.3.6 Calcium ions trigger the release of neurotransmitters from the axon terminal.

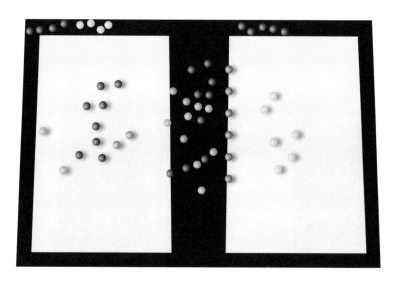

9 The neurotransmitters diffuse across the synapse and bind to the receptors on the receiving neuron, which allow sodium ions from the synapse inside (Figure 10.3.7).

FIGURE 10.3.7 Neurotransmitters open sodium ion channels that cause a new impulse in the receiving neuron.

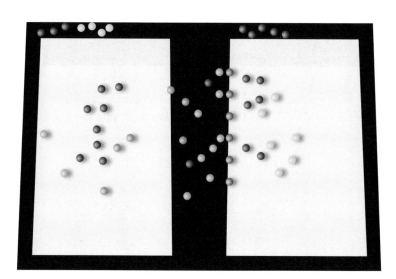

10 The receiving neuron now has an action potential where the sodium entered from the synapse. Move the action potential down the receiving neuron to complete the process.

11 Combine your photographs with movie editing software to create a rough time-lapse model of the transmission and transduction of a nerve impulse.

CHALLENGE
Modify this model to incorporate the roles of the myelin sheath and nodes of Ranvier in signal transmission.

10.4 Transmission and transduction

1 Draw a labelled diagram of the synapse between two neurons.

2 Compare and contrast transmission and transduction using the Venn diagram in Figure 10.4.1.

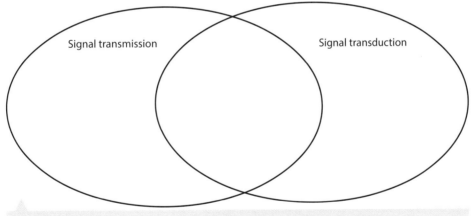

FIGURE 10.4.1 Comparing and contrasting signal transmission and signal transduction

3 Suggest why transduction is used in nerve signalling, even though transmission is considerably faster.

_____

_____

_____

_____

_____

1 Schwann cells:

    A insulate the axon

    B nourish the neuron

    C inhibit the signal

    D detect the stimulus.

2 The synapse is the:

    A cell body

    B gap between nodes

    C neurotransmitter

    D gap between neurons.

3 List the three types of neurons and their sequence in a response to stimulus.

_____

_____

_____

4 Explain the difference between axon terminals and dendrites.

_____

_____

5 Briefly outline the process of signal transmission, including the role of the nodes of Ranvier.

_____

_____

_____

6 Compare and contrast a sensory neuron with a motor neuron.

_____

_____

_____

_____

_____

9780170411660

LEARNING

Summary

▶ Hormones are chemical messengers (produced mostly in endocrine glands) that relay messages to target cells via the circulatory or lymphatic system.

▶ Cells display specific receptors for each hormone that they respond to. A cell's sensitivity to a hormone is directly related to the number of receptors it displays either on the cell surface (water-soluble hormones) or internally (fat-soluble hormones).

▶ Most cells do not maintain large numbers of receptors on the cell surface. Instead, they maintain smaller receptor populations that trigger upregulation (the production of more receptors) as well as their intended cellular response.

▶ Upregulation results in a cell becoming more sensitive to a particular hormone, while downregulation results in it becoming less sensitive to a particular hormone.

▶ When a hormone binds to its receptor, it activates a signal transduction mechanism that alters cellular activity (usually as well as producing more of the receptor).

▶ Fat-soluble hormones can pass through the cell membrane freely and bind to receptors within the cell.

▶ Water-soluble hormones cannot pass through the cell membrane and must bind to external receptors on the cell surface. The receptor then activates secondary messengers within the cell to alter cellular activity on behalf of the hormone.

11.1 Hormones as chemical messengers

Hormones are chemicals produced by the body to prompt a response from specific cells. There are more than 50 different hormones produced by the human body. While each hormone is made to target and activate a particular cell and cause a particular response, most hormones target multiple types of cells and cause different responses in each.

QUESTIONS

1. Complete the table by matching each hormone with its gland and its target effect. The first one has been done for you.

GLAND	HORMONE	TARGET EFFECT
Testes	Antidiuretic hormone	Uptake of glucose from the blood
Pancreatic alpha cells	Adrenaline	Reabsorption of water in the kidney
Pituitary	Thyroxine	Increased blood pressure by vessel constriction
Adrenal	Insulin	Increased metabolic rate in almost all tissues
Pancreatic beta cells	Glucagon	Regulation of aggression and competitiveness
Thyroid	Testosterone	Release of glucose from the liver

2. Steroids are a group of hormones that are mostly hydrophobic (fat-soluble). Given what you know about the effects of steroid use in sports, select a hormone from Question **1** that may be a steroid and justify your choice.

_____

_____

_____

_____

11.2 Upregulation and downregulation

A cell's sensitivity to any particular hormone is determined by the number of receptors that it expresses for that hormone. Having more receptors makes a cell more sensitive and responsive than having fewer receptors. A cell needs to express receptors for many different hormones. Most cells only permanently express a small number of each type of receptor and produce more when the hormone is detected. This type of response is called upregulation, while a response that dampens a cellular process is called downregulation. Both types of regulation involve changing the way a cell's DNA is translated, either producing more or less of the products of a particular gene or set of genes.

QUESTIONS

1   Define:

    a   upregulation

    _____

    b   downregulation

    _____

2   Give an example of a situation where a cell may need to become more sensitive to a particular hormone.

    _____

    _____

    _____

3   Downregulation, in a more general sense, refers to any process by which gene products are suppressed. Explain how downregulation could be helpful in homeostasis.

    _____

    _____

    _____

    _____

11.3 Receptor binding

QUESTIONS

1 Label the following diagrams as either fat-soluble or water-soluble hormone binding. Describe the features of the diagram that support your choice.

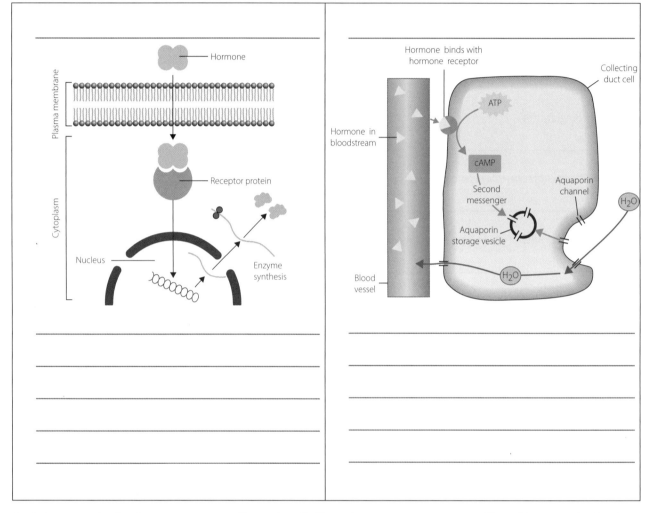

2 Substances that bind to receptors are collectively called ligands. Hormones are a type of ligand because they bind to receptors. How receptors and ligands actually bind to each other has been an active subject of study for more than 50 years and many models have been proposed to explain the process. One such model is the lock-and-key model, which has been superceded by a variant called the induced fit model.

Briefly outline the lock-and-key model and explain which of the model's limitations have been addressed in the induced fit model.

_____

_____

_____

_____

_____

9780170411660

1 Which of the following is not an endocrine gland?

   A Thyroid

   B Pancreas

   C Bladder

   D Pituitary

2 A cell's sensitivity is directly related to the:

   A number of receptors it displays

   B type of receptors it produces

   C amount of hormone it receives

   D endocrine gland it is a part of.

3 Explain the difference between the actions of fat-soluble and water-soluble hormones.

   _____

   _____

4 Define the upregulation of hormonal control pathways.

   _____

5 Explain the process of signal amplification.

   _____

   _____

   _____

   _____

6 Adrenaline is released by the adrenal glands in response to fright. Outline two different effects that this hormone has in the body and use them to explain how cell targeting works.

   _____

   _____

   _____

   _____

   _____

12 Thermoregulation

LEARNING

Summary

- Thermoregulation is the process by which animals maintain their body temperature within tolerance limits.

- The principle of thermoregulation is maintaining a rate of heat gain that equals the rate of heat loss.

- Structural adaptations, behavioural responses, physiological mechanisms and homeostatic mechanisms act to maintain body temperature.

- Structural adaptations used by endotherms to reduce heat loss include feathers and fur.

- Brown adipose tissue increases heat output.

- The shape and size of animals affects heat gain and loss. The smaller the animal, the greater the surface-area-to-volume ratio and the greater the loss of heat to the environment.

- Huddling is a behavioural response used by endotherms to reduce heat loss.

- Torpor is a state of decreased activity and metabolism. It allows animals to survive unfavourable conditions such as very high or very low temperatures by reducing metabolism and body temperature. Hibernation and aestivation are examples of torpor.

- Vasodilation is a physiological response to changes in temperature. When the temperature rises, skin blood vessels dilate (vasodilation) causing blood to flow close to the skin's surface allowing heat to leave. When the temperature drops, skin blood vessels constrict (vasoconstriction), causing reduced blood flow to the skin's surface reducing heat loss.

- Sweat glands respond to temperature increases by opening to release water and salt onto the skin. Evaporation of water from the moist skin cools the blood as it flows through capillaries near the surface.

- Counter-current heat exchange reduces heat loss by situating arteries and veins close together in extremities such as feet and fins. Heat in arteries coming from the body core is transferred directly to the returning blood in the veins rather than being lost to the environment.

- Shivering is a physiological response to cold conditions. Shivering increases heat production by increasing the cellular metabolic rate.

- Thyroxine is a hormone that increases the metabolic rate of cells, increasing heat output. The levels of thyroxine are under homeostatic control involving negative feedback.

- When metabolic rate increases, demand for glucose increases. Insulin regulates the glucose levels in blood and is under homeostatic control involving negative feedback.

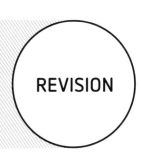
12.1 | Environmental temperature and metabolic rate

At particular external temperatures, the behaviours and physical features of an animal become inadequate and the animal cannot stabilise its body temperature. If the temperature cannot be stabilised, the metabolic rate of an animal begins to rise, which increases heat output.

Figure 12.1.1 shows the effect of environmental temperature on the metabolic rate of a generalised mammal.

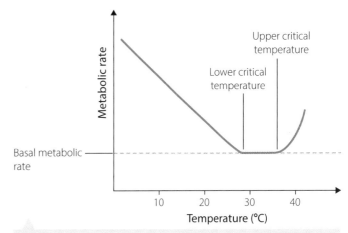

FIGURE 12.1.1 Metabolic rate of a mammal

QUESTIONS

1 At low temperatures, the metabolic rate of mammals is significantly higher than at high temperatures (Figure 12.1.1). Explain this observation.

_____

_____

2 Identify two ways a mammal could increase its metabolic rate.

_____

_____

3 The graph suggests that these animals have an optimum temperature range for body function. State that range.

_____

4 Predict what would be happening to cell function on either side of this range if the body fails to maintain a stable body temperature.

_____

_____

5 Name the process whereby some animals can survive extreme cold conditions when they are unable to maintain temperature within tolerance limits. Explain how this process works.

_____

_____

6 Name the process by which other animals can survive very hot and dry conditions.

_____

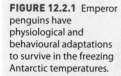

12.2 | Thermoregulation in emperor penguins

Emperor penguins are able to survive in extremely cold weather due to their ability to reduce heat loss.

FIGURE 12.2.1 Emperor penguins have physiological and behavioural adaptations to survive in the freezing Antarctic temperatures.

QUESTIONS

1 Use Figure 12.2.1 to describe how emperor penguins behave to reduce heat loss and how this behavioural adaptation increases survival rates.

_____

_____

_____

_____

2 Figure 12.2.1 shows emperor penguins in close contact with the snow and ice, yet their extremities remain warm. Describe a physiological adaptation that allows this to happen. Explain how this adaptation functions.

_____

_____

_____

_____

3 Provide an example of a structural adaptation in emperor penguins and explain how it helps maintain body temperature.

_____

_____

_____

_____

9780170411660

12.3 Strategies and adaptations for thermoregulation

Thermoregulation is the process by which animals maintain their body temperature within a normal range. The principle of thermoregulation is maintaining a rate of heat gain that equals the rate of heat loss. The structures, behaviours, physiology and homeostatic mechanisms of the animal contribute to maintaining a relatively stable body temperature (Table 12.3.1).

TABLE 12.3.1 Thermoregulation mechanisms

STRUCTURAL FEATURES	BEHAVIOURAL RESPONSES	PHYSIOLOGICAL MECHANISMS	HOMEOSTATIC MECHANISMS
Insulation	Moving locations	Vasomotor control	Hormone negative feedback
Brown adipose tissue	Huddling	Evaporative heat loss	
Shape and size	Torpor	Counter-current heat exchange	
		Thermogenesis	

Use Table 12.3.1 to provide a short description of how each thermoregulatory mechanism contributes to maintaining a relatively stable body temperature. Provide one example for each.

_____

_____

_____

_____

_____

_____

_____

_____

_____

_____

_____

_____

_____

_____

_____

12.4 Bilby thermoregulation

A group of 16 bilbies (*Macrotis lagotis*) was exposed to different temperatures and their oxygen consumption was measured. The bilbies were placed in animal chambers exposed to temperatures ranging from 5°C to 35°C.

The average results are shown in Table 12.4.1.

TABLE 12.4.1 Oxygen consumption of bilbies

TEMPERATURE (°C)	OXYGEN CONSUMPTION (mL/min)
5	2.5
10	2.0
15	1.7
20	1.4
25	0.9
30	0.7
35	0.7

1 Display the data in Table 12.4.1 as a line graph.

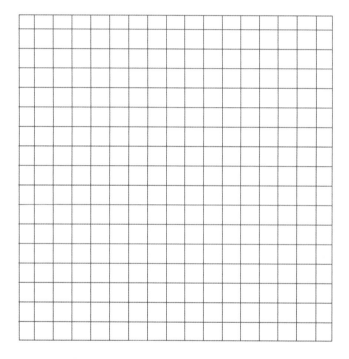

2 Explain why oxygen consumption increases as temperature decreases.

_____

_____

3 Define 'lower critical temperature' and predict what this temperature is for the bilby.

_____

_____

9780170411660

4 Suggest a reason why oxygen consumption did not vary at 30°C and 35°C.

_____

_____

5 Scientists observed the bilbies at temperatures of 5°C and 10°C and noted that they were huddling together. Explain this observation.

_____

_____

6 Identify a physiological response that would be observed at 5°C and 10°C.

_____

_____

_____

_____

_____

7 Suggest why 16 bilbies were used in the experiment rather than one bilby.

_____

_____

_____

_____

_____

EVALUATION

1 Choose the correct statement.

   A Huddling is an example of torpor.

   B Layers of feathers and fur are used to generate heat.

   C Thyroxine is secreted in response to a decrease in blood temperature.

   D Thermoregulation is the process by which animals maintain their metabolic rate.

2 Animals that live in Australian deserts have physiological and behavioural adaptations that help them survive hot conditions. Identify one physiological and one behavioural adaptation.

Physiological adaptation:

_____

Behavioural adaptation:

_____

3 **a** Name the medical emergency that occurs when a person loses heat faster than they can produce heat, causing a dangerously low body temperature.

_____

   b List two ways in which the chances of this medical emergency can be reduced.

_____

_____

4 A person moves from a heated house to the outside on a cold winter's day.

   a Name the hormone that would be secreted in response to the sudden exposure to cold conditions.

_____

   b Describe the effect of the hormone.

_____

_____

   c Describe the steps in a negative feedback mechanism by which the levels of the hormone would return to normal after moving back into the heated house.

_____

_____

_____

_____

9780170411660

5 **a** Describe how brown adipose tissue differs from white adipose tissue.

_____

_____

_____

b Explain the physiological role of brown adipose tissue in newborn humans.

_____

_____

_____

_____

_____

_____

13 Osmoregulation

LEARNING

Summary

▶ The function of osmoregulation is to control solute concentration and water balance in cells.

▶ Osmoconformers allow their osmotic concentration to be equal to the concentration of the external environment. Osmoconformers live in water that has a stable composition similar to or the same as their internal environment, so they don't lose or gain much water.

▶ Osmoregulators regulate their osmotic concentration to be either higher or lower than their external environment. They can live in environments where osmoconformers cannot.

▶ Structural features, as well as behavioural responses and physiological and homeostatic mechanisms, assist water balance maintenance in osmoregulators.

▶ A waterproof or impermeable outer layer is a structural feature that can reduce water loss.

▶ Kidneys play a key role in removing nitrogenous wastes, regulating water concentration in blood and maintaining ion levels in the blood. Kidneys have adapted to eliminate different types of nitrogenous waste (ammonia, urea and uric acid).

▶ Behavioural responses to dry conditions include burrowing.

▶ Physiological mechanisms that are used to maintain solute concentration under dry conditions include concentrating urine and tolerating higher body solute concentrations.

▶ Antidiuretic hormone (ADH) increases the permeability of cells lining the collecting ducts in kidneys, facilitating the movement of water into the surrounding blood when solute concentration is high. Negative feedback decreases the release of ADH when solute concentration decreases. This is an example of a homeostatic mechanism.

▶ Xerophytes are plants that grow under dry conditions. Structural features that reduce water loss include a thick waxy cuticle on the leaf surface, reduced numbers of stomata on the top of the leaf and increased numbers on the bottom of the leaf, sunken stomata, cylindrical or rolled leaves, reduced leaf numbers or no leaves and hairs on leaves.

▶ Xerophytes typically have long vertical roots which absorb water from deep down in the soil or superficial roots which grow out horizontally just beneath the surface to absorb water quickly before it evaporates.

▶ Mesophyte plants live in areas with adequate water. Extreme adaptations are not needed.

▶ Hydrophytes are plants that grow in or on water. As water contains much less oxygen than air, the problem for hydrophytes is lack of oxygen. Their leaves have many air-filled intercellular spaces through which air can move.

▶ Halophytes live under salty conditions. They show structural characteristics similar to xerophytes.

▶ Abscisic acid (ABA) is a hormone that homeostatically regulates water status and stomatal movement.

9780170411660

13.1 | Water balance in animals

Organisms have various mechanisms to maintain water balance. Structural features, as well as behavioural responses and physiological and homeostatic mechanisms, assist water balance maintenance in animals such as the spinifex hopping mouse (*Notomys alexis*). Spinifex hopping mice are found in desert regions in Central Australia. They can survive by eating seeds, plant roots and insects without drinking any water. Their urine is very concentrated and they eliminate faeces that is almost dry. They have a bushy end to their tails.

QUESTION

Explain how each of the adaptations of the spinifex hopping mouse shown in Figure 13.1.1 helps it maintain water balance.

Homeostatic – antidiuretic hormone (ADH)

Structural – bushy tail

Behavioural – burrowing

Structural – loop of Henle

Behavioural – nocturnal

Physiological – concentrated urine

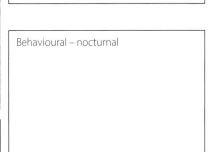

FIGURE 13.1.1 A concept map of adaptations of the spinifex hopping mouse

13.2 | Water balance in plants

Plants have a range of features that help obtain and/or retain water. Plants in different environments have adapted their leaf structure to maximise water balance.

1   Leaf cross-sections were taken from plants growing in dry terrestrial, temperate terrestrial and aquatic environments. Label the leaf cross-sections as xerophyte, hydrophyte or mesophyte.

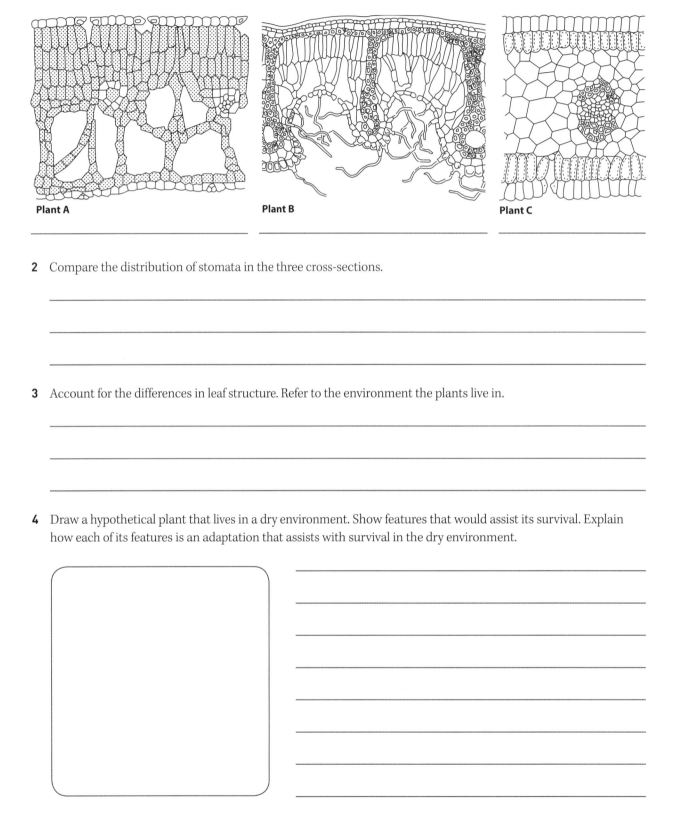

Plant A

Plant B

Plant C

_____   _____   _____

2   Compare the distribution of stomata in the three cross-sections.

_____

_____

_____

3   Account for the differences in leaf structure. Refer to the environment the plants live in.

_____

_____

_____

4   Draw a hypothetical plant that lives in a dry environment. Show features that would assist its survival. Explain how each of its features is an adaptation that assists with survival in the dry environment.

13.3 Comparing osmoregulation in different habitats

Observations were made on two mammal species in their natural habitat. Measurements were made and averages were recorded in Table 13.3.1.

TABLE 13.3.1 Observations of two mammal species

	MAMMAL SPECIES A	MAMMAL SPECIES B
Water gained through drinking (mL/day)	1000	10
Water lost through evaporation (mL/day)	150	15
Amount of urine produced (mL/day)	450	2
Level of antidiuretic hormone (ADH) (pg/mL)	1	4

QUESTIONS

1 a State which mammal species lives in a hot, dry environment.

_____

b Explain your answer using evidence from two of the measurements in Table 13.3.1.

_____

_____

_____

_____

2 The level of antidiuretic hormone varies between the two mammal species.

a Name the gland of the body that releases ADH.

_____

b Describe the effect of higher ADH levels in mammal species B compared with mammal species A.

_____

_____

_____

_____

3 Predict the difference in length of the loop of Henle in mammal species A and mammal species B.

_____

_____

_____

_____

1   Choose the option that best describes features of terrestrial plants compared with aquatic plants.

	TERRESTRIAL PLANT	AQUATIC PLANT
A	Stomata on lower surface, thick cuticle	Cuticle absent, smooth leaves
B	Stomata on upper surface, smooth leaves	Thick cuticle, large flat leaves
C	Cuticle absent, leaves with hairs	Large flat leaves, stomata on lower surface
D	Thick cuticle, large flat leaves	Leaves with hairs, stomata on upper surface

2   Most marine invertebrates are osmoconformers. Explain what this means.

_____

_____

3   Describe one adaptation found in a succulent plant to conserve water.

_____

_____

4   Describe one way a mammal could behave to reduce the amount of water lost through evaporation.

_____

_____

5   A plant leaf is examined and found to have a thick cuticle on its upper surface and sunken stomata surrounded by hairs on its lower surface.

   a   Explain how these three features assist survival of the plant in a dry environment.

   _____

   _____

   _____

   b   Describe two other features found in this plant that would assist the plant's survival in a dry environment.

   _____

   _____

6   Describe two features in an aquatic plant's leaf (e.g. water lily) structure that assist its survival in the environment it lives in.

_____

_____

9780170411660

14 Infectious disease

LEARNING

Summary

▸ Disease is any condition that interferes with the proper functioning of an organism. People differ in their susceptibility to different diseases.

▸ Infectious diseases are caused by any agent that can be transmitted from one organism to another.

▸ Non-infectious diseases are not transmitted between individuals and include nutritional diseases, genetic diseases and diseases that arise from an interaction of genetic and environmental influences.

▸ Pathogens are disease-causing agents. Cellular pathogens include bacteria, fungi, protists, endoparasites and ectoparasites. Viruses and prions are non-cellular infectious agents that are always pathogenic.

▸ Most pathogens are host-specific and each disease is characterised by its incubation period and through recognisable symptoms.

▸ Pathogens differ in their disease-causing capacity or pathogenicity.

▸ Virulence is a measure of the pathogenicity of an organism, that being the intensity of the effect of the pathogen on the host.

▸ Virulence factors are characteristics that promote the establishment and maintenance of disease. This includes the ability to stick to or invade a particular cell type, produce toxins, and cope with or avoid the host's immune system.

▸ Transmission of disease occurs by various mechanisms, including through direct contact, contact with body fluids, via contaminated food, water or disease-specific vectors.

▸ The effect of antimicrobial substances on the growth of microbiological organisms can be measured by investigation.

14.1 | Infectious and non-infectious diseases

QUESTIONS

1   Explain the difference between infectious and non-infectious diseases.

_____

_____

_____

_____

2   Indicate whether the following diseases are infectious or non-infectious by putting a tick in the appropriate place.

DESCRIPTION OF DISEASE	INFECTIOUS	NON-INFECTIOUS
Uncontrolled growth and division of lung cells		
Can be prevented by boiling water before drinking it		
Reproduction of particles inside skin cells, before escaping and entering other skin cells		
Proteins that convert the normal form of the protein to a harmful form, which can then convert more normal to abnormal forms		
Rise in blood glucose after consumption of sucrose and glucose		

9780170411660

14.2 | Viruses and bacteria

QUESTIONS

1 At each of the numbers 1–5 on Figure 14.2.1, summarise the actions of the virus as it infects a bacterial cell.

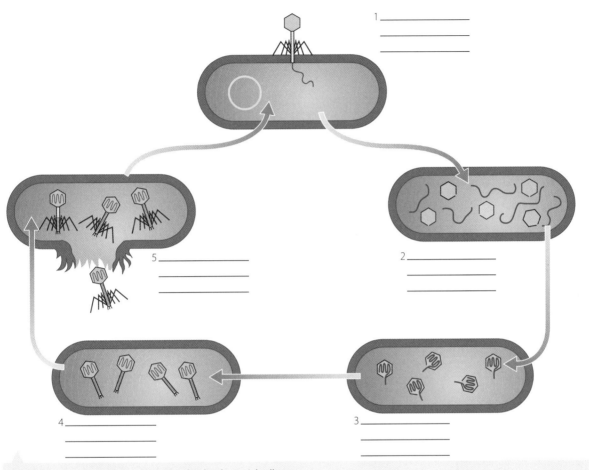

1 _____
_____
_____

2 _____
_____
_____

3 _____
_____
_____

4 _____
_____
_____

5 _____
_____
_____

FIGURE 14.2.1 Viruses reproducing inside a live bacterial cell

2 Label the structures numbered 1–8 in the bacterial cell in Figure 14.2.2.

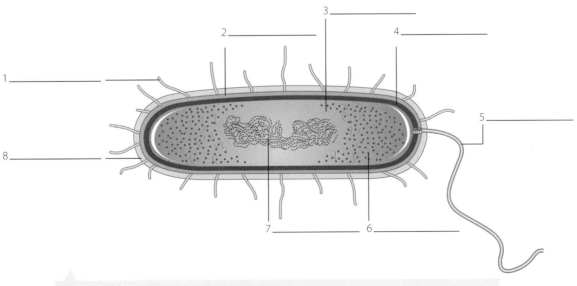

FIGURE 14.2.2 Generalised structure of a bacterium

14.3 Transmission of disease

To be able to persist and survive, pathogens must follow a repeating cycle of transmission from current to future host. Understanding their infectious cycles is critical to being able to identify suitable strategies to control pathogens.

QUESTIONS

1 Complete Table 14.3.1 for some diseases and their methods of transmission.

TABLE 14.3.1 Some diseases and their methods of transmission

DISEASE	PATHOGEN	METHOD OF TRANSMISSION
	Virus	Contact with body fluids via coughs and sneezes, and wiping nose and touching others
Tinea (athlete's foot)		Direct contact
Typhoid	Bacteria	
Bubonic plague		Insect vector – flea
Gonorrhea	Bacteria	
Rotavirus		Contaminated food
	Virus (Herpes simplex)	Direct contact
Salmonella food poisoning	Bacteria	

9780170411660

2 Consider Figure 14.3.1, which highlights the details of the transmission of malaria.

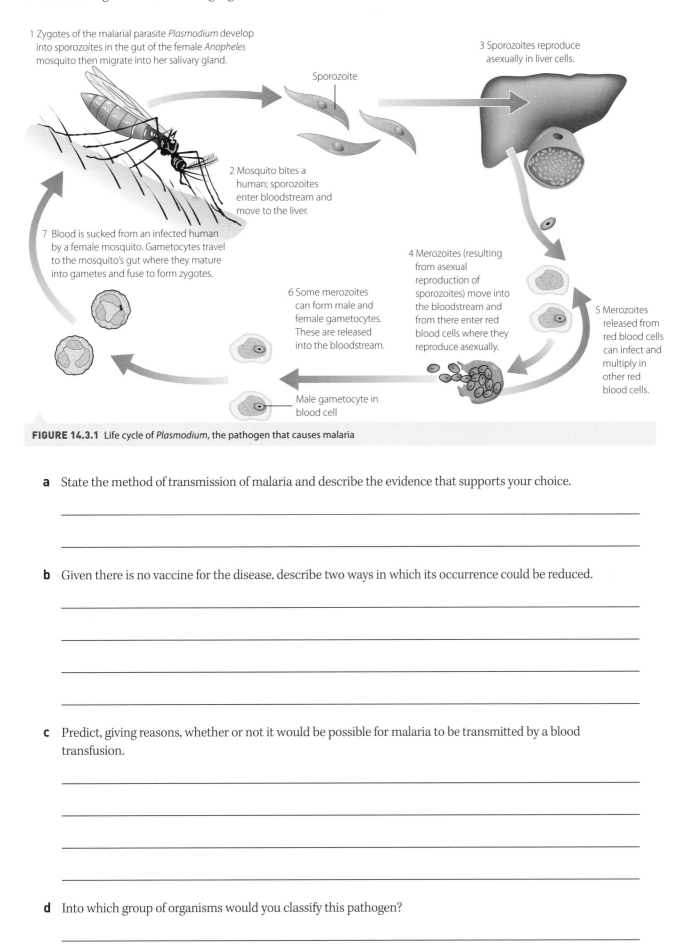

1 Zygotes of the malarial parasite *Plasmodium* develop into sporozoites in the gut of the female *Anopheles* mosquito then migrate into her salivary gland.

Sporozoite

3 Sporozoites reproduce asexually in liver cells.

2 Mosquito bites a human; sporozoites enter bloodstream and move to the liver.

7 Blood is sucked from an infected human by a female mosquito. Gametocytes travel to the mosquito's gut where they mature into gametes and fuse to form zygotes.

4 Merozoites (resulting from asexual reproduction of sporozoites) move into the bloodstream and from there enter red blood cells where they reproduce asexually.

6 Some merozoites can form male and female gametocytes. These are released into the bloodstream.

5 Merozoites released from red blood cells can infect and multiply in other red blood cells.

Male gametocyte in blood cell

FIGURE 14.3.1 Life cycle of *Plasmodium*, the pathogen that causes malaria

a State the method of transmission of malaria and describe the evidence that supports your choice.

_____

_____

b Given there is no vaccine for the disease, describe two ways in which its occurrence could be reduced.

_____

_____

_____

_____

c Predict, giving reasons, whether or not it would be possible for malaria to be transmitted by a blood transfusion.

_____

_____

_____

_____

d Into which group of organisms would you classify this pathogen?

_____

3 Consider Figure 14.3.2, which highlights the disease schistosomiasis.

FIGURE 14.3.2 Life cycle of the blood fluke *Schistosoma japonicum*, which causes schistosomiasis

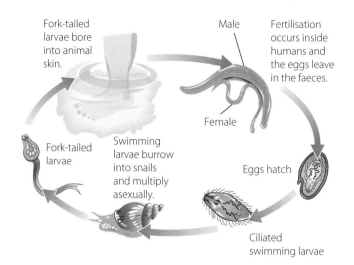

Fork-tailed larvae bore into animal skin.

Male

Fertilisation occurs inside humans and the eggs leave in the faeces.

Female

Fork-tailed larvae

Swimming larvae burrow into snails and multiply asexually.

Eggs hatch

Ciliated swimming larvae

a Describe the method of transmission and describe the evidence that supports your choice.

_____

_____

_____

b What name is given to the snail in this disease?

_____

c What name is given to the human in this disease?

_____

d Describe two ways in which the occurrence of this disease could be reduced.

_____

_____

_____

_____

e Predict the benefit to the blood fluke of having the snail involved in the life cycle.

_____

_____

9780170411660

14.4 Adaptations to ensure transmission of pathogens to new hosts

Use the answers from Table 14.3.1 on page 108 and the information provided in each question to explain the adaptations that ensure the transmission of pathogens in each scenario.

QUESTIONS

1 The cold virus irritates the mucous membranes lining the nose and throat making the host secrete large amounts of mucus, which is teeming with millions of copies of the cold virus. The host will often cough and sneeze and wipe away excess mucus with their hands.

Two adaptations for transmission are:

_____

_____

_____

_____

2 The *Herpes* virus, transmitted by direct contact, may cause asymptomatic shedding of the virus on the skin between the occurrence of visible sores. Visible sores are also very itchy.

Two adaptations for transmission are:

_____

_____

_____

_____

3 Rotavirus gastroenteritis is spread to food from the faeces of an infected person. A person with the disease has symptoms of messy, watery diarrhoea which may contain up to 10 000 million (10^{10}) virus particles per millilitre of faeces. The infective dose is only 100–10 000 virus particles.

An adaptation for transmission is:

_____

_____

4 When mature, the adult female the human pinworm nematode *Enterobius vermicularis* moves down the large intestine and exits the host via their anus to lay a batch of eggs on the surrounding skin. She then dies. The eggs cause intense itching, especially at night.

An adaptation for transmission is:

_____

_____

5 Bubonic plague is a bacterial disease of rodents caused by *Yersinia pestis* that can be spread to humans and other animals by infected rat fleas. The bacteria multiply inside the flea, migrating to the flea's stomach where they stick together to form a plug. As the plug cannot be digested, starvation induces the flea to voraciously bite its host, which dies quickly. The flea then searches for other mammals to feed from. When it bites them, it vomits blood tainted with the bacteria back into the bite wound and passing on the disease.

Two adaptations for transmission are:

_____

_____

_____

_____

6 *Salmonella* bacteria are able to reproduce in and on food they contaminate. The infective dose for *Salmonella* is $>10^5$ (100 000).

An adaptation for transmission is:

_____

_____

14.5 | Virulence factors

Only a small proportion of micro-organisms are pathogens. Virulence factors aid pathogenesis.

QUESTIONS

1 Define:

a pathogenicity

_____

b virulence

_____

2 Explain what is meant by virulence factors.

_____

_____

_____

_____

9780170411660

3 Complete the table to compare and contrast exotoxins and endotoxins.

CHARACTERISTIC	EXOTOXIN	ENDOTOXIN
Relative toxicity		
Chemical make-up		
Source		
Nature of their action		
End result		

4 Fill in the missing words in this cloze activity.

Virulence factors that facilitate _____ (binding to host cell surfaces) are known as _____. Adherence factors ensure that the pathogen attaches to the _____ type in which it can survive. Many pathogenic bacteria recognise and attach to epithelial surfaces by using _____. These are fine filaments of _____, up to several _____ in length and resembling fine hairs. Adhesins also include a wide variety of other _____ proteins, as well as bacterial cell _____ and bacterial _____.

Invasion factors are virulence factors that facilitate bacterial _____ of a host. They play a role in enabling _____ into the cells and tissues of the host in order to ensure its _____. Invasion factors are often _____ secreted by bacteria. One such _____ degrades a structural component of _____ clots, facilitating bacterial transport across _____ layers and penetration into the _____ tissues. A successful invasion by intracellular pathogens means penetrating host cell _____. Surface proteins found on some _____ allow them to invade mammalian cells via _____ proteins.

Lifestyle changes, for example _____ formation, can result in increased _____. The _____ that causes tetanus can last for years in soil as an _____ spore that will resume growth when conditions become more favourable inside a new host.

The _____ is a large, well-organised layer made of thick _____ gel that forms part of the outer _____ of many bacterial cells. Encapsulated strains of _____ have been shown to be more _____ than non-encapsulated strains, apparently because capsules inhibit _____ by host phagocytes.

A group of students completed the following practical investigation.

INTRODUCTION

Tears contain a powerful substance called lysozyme, which kills bacteria before they are able to infect our eyes. The effectiveness of lysozyme can be compared with other antibacterial substances, such as antiseptics and disinfectants.

AIM

To compare the antibacterial effectiveness of lysozyme from tears with an antiseptic and a disinfectant

MATERIALS AND METHOD

1 Spread 1 mL of bacterial broth evenly over each of three pre-made agar plates.

2 Use a marking pen to divide the base of the plate into four quarters, labelling each quarter: water, lysozyme, antiseptic and disinfectant.

3 Make 10 mL of disinfectant and antiseptic solutions by diluting with distilled water according to directions and fill another 10 mL beaker with distilled water.

4 Make 12 circles of filter paper with a hole punch.

5 Obtain tears by holding a cut onion near one eye and blinking to get tears to run down the face.

6 One at a time, dip three circles of paper into the tears, then gently place one circle onto each of the three agar plates in the correct quadrant. Repeat for each liquid, until all 12 circles have been dipped and placed on the agar.

7 Seal the plates with sticky tape and incubate them at 25°C for 24 hours.

8 Ensure the bench is wiped down with bleach and wash your hands thoroughly.

9 The next day, use a ruler to measure the diameter of the zone of inhibition, which is the clear area around each disc. This shows the degree of sensitivity of the bacteria to each substance.

RESULTS

Table 14.6.1 shows the data that one group of students obtained after following the above method.

TABLE 14.6.1 Student results

| TRIAL | DIAMETER OF ZONE OF INHIBITION (mm) FOR EACH SUBSTANCE | | | |
	LYSOZYME	ANTISEPTIC	DISINFECTANT	WATER
1	11	13	15	6
2	16	17	13	8
3	12	12	17	7
Mean				

ANALYSIS OF METHOD

1 Identify two steps in the method that were used to ensure the safety of the students and explain why each was necessary.

_____

_____

_____

_____

2 Explain the role of the disc dipped in water.

_____

_____

3 Explain the reason for using three agar plates.

_____

_____

ANALYSIS OF RESULTS AND DISCUSSION

4 Calculate the mean values for each treatment and add them to Table 14.6.1.

5 Identify the independent and dependent variables in this practical.

_____

_____

6 Describe the results by stating the order of effectiveness of each of the three solutions as bactericides.

_____

_____

7 Comment on the variation seen in the results for the three trials of each substance.

_____

_____

1 Bovine spongiform encephalopathy is caused by a:

A bacterium

B protist

C prion

D virus.

2 Use Figure 14.7.1 to analyse details of the transmission of African sleeping sickness.

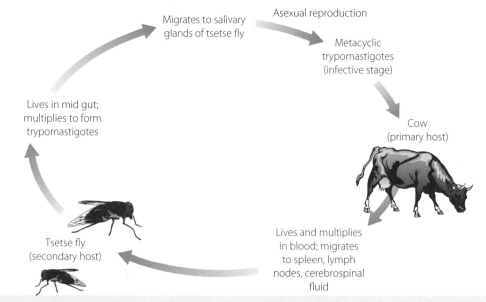

FIGURE 14.7.1 Life cycle of *Trypanosoma brucei*, which causes African sleeping sickness

a State the method of transmission and explain the benefit of this mode of transmission.

_____

_____

_____

b Describe two practical ways in which the occurrence of this disease could be reduced.

_____

_____

_____

_____

c Predict the benefit of this parasite having both sexual and asexual reproduction.

_____

_____

3 Compare and contrast endotoxins and exotoxins by describing two ways in which they are similar and two ways in which they are different.

_____

_____

_____

_____

_____

_____

_____

_____

15 Immune response and defence against disease

LEARNING

Summary

◗ To defend itself from invasion by pathogens, an organism must be able to distinguish between self (belonging in that organism) and non-self (foreign).

◗ Both plants and animals can be alerted to the presence of pathogens by detecting foreign chemicals both on the pathogen's surface and in toxins it produces.

◗ Plants and animals have innate immune defences that detect and respond to any invader, regardless of its type.

◗ Vertebrates have adaptive immune responses that develop as a result of exposure to a specific pathogen.

◗ In plants, innate immune responses to infection by a pathogen include chemical and physical defences.

◗ Surface barriers in vertebrates use chemical and physical means to prevent the entry of pathogens. Two other key weapons of the innate immune response in vertebrates are inflammation and the complement system.

◗ The adaptive immune system is able to recognise a wide range of antigens and respond specifically to different antigens. It also displays memory. These two features distinguish it from the innate immune system.

◗ The binding of an antigen causes a lymphocyte to rapidly divide to produce effector and memory cells; a process known as clonal selection.

◗ There are three major groups of lymphocytes, B cells, cytotoxic T lymphocytes (T_C cells) and helper T cells (T_H cells).

◗ Antibodies produced by plasma B cells can lead to the destruction of pathogens in several ways: agglutination, opsonisation, neutralisation and complement activation.

◗ Cell-mediated immunity involves the destruction of virally infected or cancerous cells by T_C cells, as well as the rejection of transplanted organs.

◗ Both passive and active immunity are important factors in human and animal health and vaccination is an important tool in preventing the spread of disease.

9780170411660

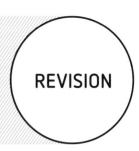

15.1 | Detection of invaders

Before a pathogen can be attacked and expelled, it must be detected by the host.

QUESTIONS

1 Distinguish between 'self' and 'non-self'.

_____

_____

2 Define 'antigen'.

_____

3 Fill in the missing words in this cloze activity.

Both plants and animals are alerted to the invasion of bacteria and viruses by _____ and _____ changes that occur in their cells or tissues. Antigens are generally protein or polysaccharide molecules _____ to the host. Their presence, either on the outer _____ of the invaders or in the _____ and enzymes they secrete, stimulates host _____ responses that usually lead to the _____ and removal of the pathogen.

Pattern _____ receptors are proteins used by nearly all organisms to identify molecules associated with pathogens. These _____ are commonly found on the _____ and in the _____ of host body cells. They recognise specific _____ and _____ patterns that are characteristic of a wide _____ of pathogens, but not found on _____ cells. These molecules are called pathogen-associated molecular patterns (_____) and include lipopolysaccharides, glycoproteins and particular protein sequences on the _____ of invaders. Even a small part of a molecule, called an _____, may be antigenic.

A particular receptor can recognise a variety of different pathogens if all of them display the same _____ pattern. For example, the material that makes up bacterial flagella, called _____, is found in a wide variety of bacteria. This enables a _____ receptor to recognise many different types of _____ as invaders. This system of _____ has the advantage of activating a _____ response to invaders but it lacks a high degree of _____.

15.2 | Innate and adaptive immune responses

The innate and adaptive immune responses are closely linked: the detection of invaders and subsequent initiation of an innate immune response is required for an adaptive immune response to occur.

ACTIVITY 1

Both plants and animals have a variety of barriers to prevent the entry of pathogens.

1 At each of the numbers 1–10 on Figure 15.2.1, summarise the physical and chemical barriers to infection in humans.

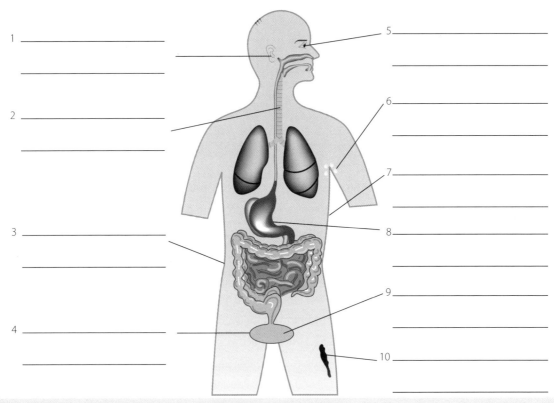

FIGURE 15.2.1 Summary of the physical and chemical barriers to pathogenic infections in a human

9780170411660

2 At each of the numbers 1–3 on Figure 15.2.2, summarise the barriers to pathogenic infections found in plants.

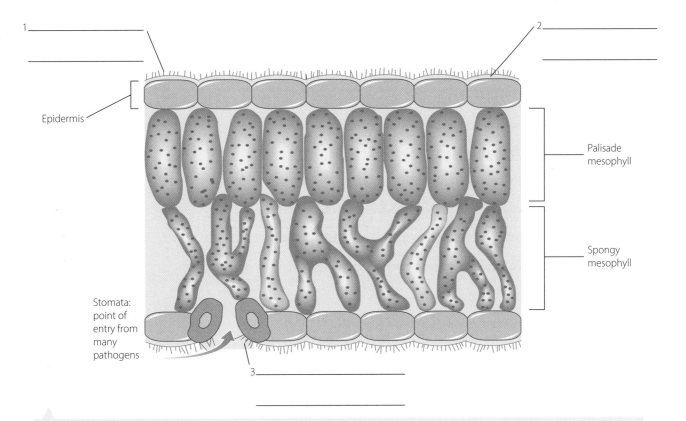

1 _____

_____

2 _____

_____

Epidermis

Palisade
mesophyll

Spongy
mesophyll

Stomata:
point of
entry from
many
pathogens

3 _____

_____

FIGURE 15.2.2 A cross-section of a typical dicotyledon leaf showing some barriers to pathogens found in plants

ACTIVITY 2

Circle the correct response: true or false.

1 Adaptive responses are more rapid than innate responses. True/False

2 Innate responses to infection have ancient origins and are highly effective. True/False

3 Adaptive responses exist only in vertebrates. True/False

4 Retaining immunological memory from previous experience of a specific pathogen is an important feature of innate immunity. True/False

5 The adaptive responses of plants, invertebrates and mammals are remarkably similar. True/False

6 Innate immune responses are general and non-specific. True/False

7 Adaptive immune responses form the natural resistance with which any organism is born. True/False

8 Because of the actions of the innate immune system, people surviving diseases such as smallpox seldom contract the disease again. True/False

9 Adaptive immune responses are highly specific because they attack only the pathogen that stimulated the response. True/False

10 Innate immune responses are termed non-specific because they are capable of changing the organism after it experiences the pathogen. True/False

15.3 | Response of a body under attack

Once an invader is identified, the immune system's response is rapid and effective.

QUESTIONS

1 Use the numbers 1–11 to indicate the order in which these actions would occur in a person.

   a Complement proteins cause lysis of bacterial cells.   _____

   b Phagocytes engulf bacterium destroyed by antibodies.   _____

   c Histamines secreted by mast cells cause inflammation.   _____

   d Antigens from bacterium bind with antibody on immature B cell.   _____

   e T-helper cell secrets lymphokines to help B cells divide.   _____

   f B cell divides to produce many plasma and memory cells.   _____

   g Plasma cells produce many antibodies.   _____

   h Antibodies bind with antigens on bacterium.   _____

   i Mucus membrane lining trachea traps bacteria.   _____

   j Memory cells migrate to lymph nodes ready to fight a second infection.   _____

   k Cilia lining trachea move mucus up and out of the body.   _____

2 Identify which of the options in Question **1** are innate responses.   _____

15.4 | Actions of lymphocytes

The adaptive immune response involves activation of specific immune cells called lymphocytes.

ACTIVITY 1

1 Complete the table to compare B lymphocytes, helper T lymphocytes and cytotoxic T lymphocytes.

	B LYMPHOCYTES	HELPER T (T_H) LYMPHOCYTES	CYTOTOXIC T (T_C) LYMPHOCYTES
Site of development of self-tolerance			
Undergo clonal selection (Yes/No)			
Effector functions			
Formation of memory cells (Yes/No)			

2 The binding of antibodies to antigen can cause the destruction of pathogens in four ways. Describe each of these.

   a Complement activation:

   _____

   _____

b Opsonisation:

_____

_____

c Neutralisation:

_____

_____

d Agglutination:

_____

_____

ACTIVITY 2

Liver, heart and kidney transplants are now fairly common procedures in many hospitals. However, transplant patients face the problem of rejection of these organs.

1 Explain why the immune system rejects these organs.

_____

_____

Transplant patients are usually prescribed immunosuppressant drugs to prevent transplant rejection. Many immunosuppressant drugs work by interfering with DNA synthesis.

2 Suggest a negative effect that these drugs may have on the health of the patient.

_____

_____

3 Explain how a drug that interferes with DNA synthesis can prevent transplant rejection.

_____

_____

_____

_____

4 A patient with kidney failure was successfully treated with a kidney transplant from his identical twin brother. He was concerned that the doctor did not prescribe immunosuppressant drugs. Are the patient's fears warranted? Justify your response.

_____

_____

_____

15.5 | Antibodies in action

Antibodies are the key to both active and passive immunity. Commercially produced antibodies can be used to treat snakebite.

ACTIVITY 1

Complete the table to demonstrate how active and passive immunity can be acquired.

	ACTIVE IMMUNITY	PASSIVE IMMUNITY
Naturally occurring	Exposure to a pathogen	1
		2
Artificial		1
		2

ACTIVITY 2

Australia has many venomous snakes. One species, commonly called the death adder (*Acanthophis antarcticus*), has one of the most dangerous bites in the world. Fortunately, there is an anti-venom available for people who have been bitten by a death adder. If the anti-venom is injected quickly enough, it prevents paralysis and death.

Anti-venom is prepared by injecting tiny amounts of snake venom into a horse over a long period of time. The amounts of venom injected are so small that the horse is unaffected; however, there is a response by the horse's immune system.

1 Name the substances the horse would produce in order to counteract the snake venom in its body.

_____

2 Name the cells in the horse that would be responsible for the formation of this substance.

_____

3 Outline the steps involved in the formation of these substances.

_____

_____

_____

_____

_____

_____

4 Explain why small amounts of venom are injected into the horse over a long period of time.

_____

_____

_____

5 After 10–12 months, blood is extracted from the horse and the plasma can be injected into snakebite victims. Identify the term given to the use of horse plasma as a treatment for snakebite.

_____

6 Explain how this is effective in treating the snakebite victim.

_____

_____

_____

EVALUATION

1 Invasion by a pathogen can have several consequences. Choose the correct alternative.

   A Inflammation involves the detection of a foreign invader and subsequent activation of cells of the immune system, particularly phagocytes.

   B The complement cascade can result in the formation of the membrane attack complex in the lymphocytes' cell membranes, causing osmotic cell lysis.

   C Cytotoxins are secreted by activated cells during infection and stimulate local inflammatory responses, including recruitment of cells to sites of infection.

   D Phagocytes are an essential component of the adaptive immune system, clearing pathogens and apoptotic cells.

2 Plants have important chemical defences against invaders.

   a Name and describe the defensive action of two plant toxins.

   _____

   _____

   _____

   _____

   b State the chemical nature of plant defensins and describe two ways in which they protect plants from pathogens.

   _____

   _____

   _____

   _____

3 Cigarette smoke has been shown to decrease ciliary beat frequency and reduce the number of ciliated cells in the airway epithelium. Predict two effects of smoking on the body's defences.

   _____

   _____

   _____

   _____

9780170411660

4 A person needs clotting proteins in their blood to form a blood clot; normally, the liver uses vitamin K to make these proteins. Vitamin K is obtained from many foods, especially green vegetables. Warfarin is a drug that reduces the liver's ability to use vitamin K to make these blood-clotting proteins.

a Warfarin has been called a blood thinner. Create an argument that either refutes or supports this name.

_____

_____

_____

_____

b Predict the effect on a person of a diet very low in green vegetables.

_____

_____

_____

5 Immune thrombocytopaenic purpure (ITP) is an autoimmune disease in which the platelet counts drop to extremely low levels. Patients may develop bruising, rashes and, in extreme cases, severe internal bleeding. Antibodies against platelet surface markers can often be found in the bloodstream of patients with ITP.

a Describe the role of platelets.

_____

_____

b Explain how the formation of anti-platelet antibodies may lead to the symptoms described.

_____

_____

_____

c Platelets from blood donations can be given to patients as a transfusion. Predict whether or not these would be effective at preventing symptoms in patients with ITP. Justify your response.

_____

_____

_____

16 Transmission and spread of disease

Summary

▶ Movement of individuals can help spread pathogens into new populations. Air travel has increased the ease with which individuals can move between populations and can allow for rapid spread of disease.

▶ The transmission of infectious disease is affected by characteristics of each disease, environmental factors and features of the affected population. These include mechanism of transmission, persistence in host, population density, host movements and the proportion of the population that is immune.

▶ Quarantine is a measure used to prevent the spread of disease into healthy populations. In Australia, the emphasis of quarantine policy has changed over time from human diseases to plant and animal pathogens that threaten our unique wildlife.

▶ It can be challenging to understand and predict the transmission of any particular disease because there are so many factors that influence this process.

▶ Human activities can have substantial impacts on the transmission of disease. These impacts may be intended (as in the case of public health measures) or unintended.

▶ Hand hygiene, travel restrictions, school and workplace closures, reducing large gatherings of people, temperature screening, contact tracing and quarantine are strategies that can help prevent the spread of disease.

▶ Immunisation is an important tool in preventing the spread of disease to individuals and throughout a population (through herd immunity).

▶ Systems for monitoring the spread of disease are needed so that public health interventions can be well targeted. The most common form of disease monitoring involves reporting of notifiable diseases.

▶ Management of an outbreak of a disease requires an investigation to discover the causative factors.

9780170411660

16.1 | Describing disease

Communication is vital in controlling the transmission and spread of disease.

1 Use your knowledge of disease to write a short description of an infectious disease.

_____

_____

_____

_____

2 Complete the table to match the concepts with their definitions by choosing from the word list.

> Endemic   Infectivity   Epidemic   Morbidity
> Virulence   Mortality   Outbreak

CONCEPT	DEFINITION
	A very serious increase in the occurrence of a particular disease above the baseline level for that population
	A disease that is prevalent at a constant rate within a population
	The impact of a disease within a population, measured by the number of people affected by that disease
	An increase in the occurrence of a particular disease above the baseline level for that population
	The impact of a disease within a population, measured by the number of deaths caused by that disease
	The ability of a pathogen to cause severe disease within its host
	The ability of a pathogen to spread from one host and infect another host

16.2 | Analysing the consequences of infection

Figure 16.2.1 shows that infection with a pathogen has different effects in different people, depending on factors such as their susceptibility and levels of immunity.

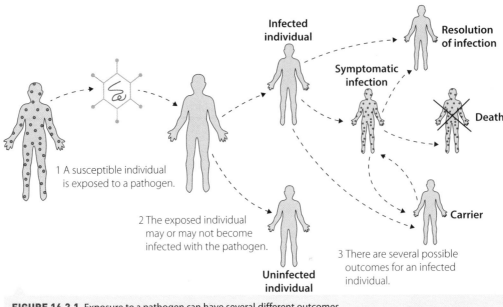

FIGURE 16.2.1 Exposure to a pathogen can have several different outcomes.

1 Name the type of pathogen in this scenario.

_____

2 List five modes of transmission by which this pathogen could be passed from one person to another.

_____

_____

3 Describe two factors that would affect whether or not the susceptible individual would become infected with the pathogen.

_____

_____

4 Describe the events that would have occurred to give rise to each of the four outcomes shown for an infected person.

_____

_____

_____

_____

5 Give an example of a disease that can remain in asymptomatic carrier individuals for a long period of time.

_____

9780170411660

6 Explain why asymptomatic carriers present a significant problem in the control of disease.

_____

_____

_____

16.3 | Case study: Spanish flu

As World War I drew to a close, the 1918–19 Spanish influenza outbreak spread rapidly throughout the world, killing up to 100 million people, including about 12 000 Australians. The reason this outbreak was so devastating illustrates the many factors involved in the spread of any disease.

The fact that populations were exhausted, living in close quarters with poor hygiene and poorly nourished as a result of the war increased susceptibility to infection. The transportation of soldiers around the world was also an important factor. War had also disrupted normal healthcare programs, leaving countries unprepared to respond. Additionally, this particular strain of influenza was more virulent than regular seasonal influenza and had high infectivity. It especially affected people with strong immune systems and it is now thought that an overreaction of many people's immune systems led to their deaths.

1 Define:

a virulent

_____

b infectivity

_____

c susceptibility.

_____

2 Explain why exhaustion, poor nourishment, living in close quarters and poor hygiene caused increased susceptibility to infection.

_____

_____

3 Spanish influenza provides a strong example of how many factors contribute to the spread of disease. Classify the contributing factors reported in the article as relating to the:

a host

_____

_____

_____

b pathogen

_____

_____

_____

c environment

_____

_____

_____

4 The overreaction of a person's immune system to produce a cytokine storm in the lungs can occur when the immune system encounters a new and highly pathogenic invader. Use your knowledge of the immune system to suggest why this reaction causes death.

_____

_____

_____

_____

5 Provide another example of a disease that has rapidly spread around the world because of international travel.

_____

_____

_____

16.4 Defining disease

Draw lines to join the concepts to their correct descriptions.

CONCEPT	DEFINITION
Quarantine	The study of the causes and effects of diseases at a population level
Intermediate host	An infected person able to control a disease to some degree, but still capable of transmitting infection to others
Isolation	A host in which the adult phase of a parasite produces gametes
Definitive host	The enforced isolation of individuals at risk of carrying disease to prevent the spread of that disease into healthy populations
Carrier	A living organism that transmits pathogens from one host to another
Vector	An organism in which a pathogen or parasite undergoes development and spends a small proportion of its life cycle
Epidemiology	The enforced separation of individuals with a disease to prevent the spread of that disease into healthy populations

16.5 Outbreak investigation

An outbreak investigation involves a series of steps that aim to determine what has caused the outbreak of disease. Fill in the missing words in these cloze activities.

1 The first step is to confirm that the reported cases meet the definition of an _____. This involves comparing the number of diagnosed cases with _____ levels of the disease.

2 Next, investigators formulate a _____ definition. Case definitions include the type of illness, the place and the time.

3 Investigators then find people affected by the outbreak by using _____ tracing; recent _____ of the _____ person are traced and screened for the infection.

4 The type of _____ sought will vary with the _____ of transmission. In a case of food poisoning, investigators focus on people _____ to the same food sources, not direct _____ with an infected individual.

5 Investigators then gather _____ in interviews in which they ask about usual activities, sick contacts, recent meals and travel.

6 Once a _____ about how the outbreak is spreading has been generated, the investigators search for _____ to support or refute that hypothesis.

7 In a classic case of food poisoning, in a small coastal town in Australia, health authorities were alerted to a potential outbreak of gastroenteritis among people who had eaten at a particular restaurant. All of those who had dined at the restaurant on that night were contacted and asked about symptoms. A total of 19 people contracted gastroenteritis.

One of the patients was tested and found to be infected with norovirus, a common cause of gastroenteritis. Investigators collected detailed information about what each person ate, as shown in Table 16.5.1. No evidence of the virus was found in the kitchen.

TABLE 16.5.1 Items eaten by those who did and did not contract gastroenteritis following a dinner function

	CUSTOMERS WHO ATE THE FOOD		CUSTOMERS WHO DID NOT EAT THE FOOD	
	CONTRACTED INFECTION	TOTAL	CONTRACTED INFECTION	TOTAL
ENTRÉE				
Oysters	19	34	0	19
Prawns	17	37	2	16
Lettuce garnish	10	15	9	38
Cocktail sauce	11	20	8	33
Chicken skewers	1	16	18	37
MAIN				
Leg ham	17	40	2	13
Lamb	15	42	4	11
Beef	11	41	8	12
Chicken	12	37	7	16
Cucumber and tomato salad	5	7	14	46
DESSERT				
Pavlova	6	9	13	44
Toffee pudding	0	6	19	47
Apple strudel	3	6	16	47
Fruit salad	3	8	16	45
Cream	11	25	8	28

Source: Huppatz, C., Munoch, S.A., Worgan, T., Merritt, T.D. et al. (2008) 'A norovirus outbreak associated with consumption of NSW oysters: implications for quality assurance systems', *Communicable Diseases Intelligence*, 32(1), pp. 88–91.

a State the case definition for this outbreak.

_____

_____

b Define 'contact tracing'.

_____

_____

c Describe the contact tracing in this scenario.

_____

_____

d Using the data in Table 16.5.1 construct a hypothesis, supported by evidence, to explain this outbreak.

_____

_____

_____

_____

9780170411660

Traditionally, disease activity has been monitored through the notification of public authorities after diagnosis by a doctor. Recently, another method of disease surveillance, using Google search engine data, has been investigated. The graphs in Figure 16.6.1 show how the pattern of Google searches on flu-like symptoms relates to the number of confirmed cases of flu.

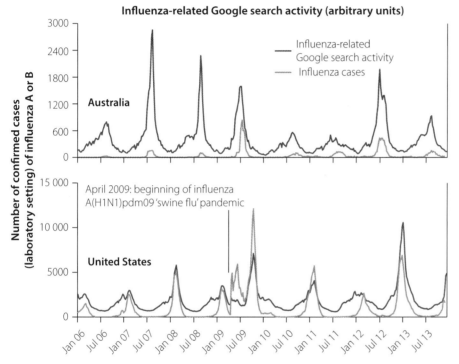

FIGURE 16.6.1 Patterns of influenza infection and Google search activity related to flu-like illness in the United States and Australia

Sources of data: Google FluTrends (http://www.google.org/flutrends) and WHO FluNet

1   Use your knowledge of disease, the information above and the graphs in Figure 16.6.1 to choose the *incorrect* alternative.

   A One limitation of the new method is that the number of influenza cases in the US at the beginning of the pandemic is greater than those predicted by the Google search data.

   B One advantage of the new method is that peaks in the Google search data correspond with peaks in the number of influenza cases being diagnosed.

   C One limitation of the new method is how effective algorithms are at determining whether a search is actually about an illness that a person is experiencing.

   D One advantage of the new method is that it provides information to health authorities in real time.

2 Explain why the movement of individuals and populations can facilitate the spread of disease.

_____

_____

3 State the name given to a scientist who studies the causes and effects of diseases at a population level.

_____

4 Malaria is a disease caused by protists from the _Plasmodium_ genus and transmitted between human hosts by the _Anopheles_ mosquito. Malaria is found only in areas of South America, Africa and Asia that are near the equator, because that species of mosquito can only live in these areas.

   a State a word that describes the role of the _Anopheles_ mosquito in the spread of malaria.

   _____

   b Explain why scientists fear that malaria could become more widespread with global warming.

   _____

   _____

5 By referring to Figure 16.6.2, describe herd immunity and explain its importance.

_____

_____

_____

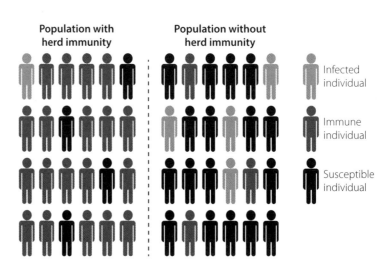

FIGURE 16.6.2 Herd immunity

Population with herd immunity

Population without herd immunity

Infected individual

Immune individual

Susceptible individual

9780170411660

6 Scientists have developed many strategies to control the spread of disease. Explain how each of the following prevent the spread of disease.

a School and workplace closures:

_____

_____

b Reduction of mass gatherings:

_____

_____

c Temperature screening:

_____

_____

d Travel restrictions:

_____

_____

e Contact tracing:

_____

_____

f Quarantine:

_____

_____

g Personal hygiene measures:

_____

_____

BIOLOGY UNITS 1 & 2

MULTIPLE-CHOICE QUESTIONS

Clearly indicate ONE response for each multiple-choice question. No marks will be deducted for incorrect responses.

Question 1

Which statement correctly describes the role of proteins embedded in the phospholipid bilayer?

A Transport proteins act as passageways that allow specific substances to move across the membrane.

B Carrier proteins within membranes assist other molecules to cross the membrane in active transport, but not facilitated diffusion.

C A channel protein is a type of transport protein that forms narrow passageways to allow the passage of hydrophobic substances across the membrane with the use of energy.

D One type of membrane transport protein that effectively acts as a one-way valve does not require energy to move molecules or ions up their concentration gradient.

Question 2

Advantages in an organism having membrane-bound organelles *do not* include a membrane-bound organelle being able to:

A confine important but harmful molecules, protecting the rest of the cell from their harmful effects

B carry out specific functions without interfering with other activities elsewhere in the cell

C facilitate the synthesis or the breakdown of complex molecules

D be localised in particular regions of the cytoplasm where they are needed.

Question 3

When an apple is cut, an enzyme causes compounds in the exposed tissue of the apple to react rapidly with oxygen. This starts a process that makes the apple go brown. If apple pieces are plunged into boiling water for 5 minutes, then cooled, the browning is slowed because:

A the substrate is denatured by the boiling water

B the enzyme permanently changes shape

C the enzyme is washed off the apple in the water

D boiling water enters the apple cells by osmosis, causing them to burst.

Question 4

In the process of:

A aerobic respiration, plants convert light energy into chemical energy

B aerobic respiration, animals but not plants convert chemical energy into heat energy

C photosynthesis, plants but not animals convert chemical energy into heat energy

D photosynthesis, plants convert light energy into chemical energy.

Question 5

Which of the following statements about the removal of carbon dioxide from the blood is correct?

A The constant movement of blood maintains a concentration gradient, ensuring carbon dioxide remains higher in the blood than in the alveolus.

B The alveolar wall and the capillary wall are each one cell thick to slow movement of carbon dioxide into the alveolus.

C The alveolar wall is only one cell thick to increase the surface area of lung available for removal of carbon dioxide.

D Some carbon dioxide is moved from the blood into the alveolus by active transport.

Question 6

Some actions that occur in the nephron of a healthy human kidney are listed below.

1 Selective reabsorption of glucose from the filtrate into the blood

2 Water flows out of the tubule and into the blood by osmosis

3 Secretion of wastes from the capillary network into the duct

4 Movement of water, urea, glucose, amino acids, and salt into Bowman's capsule by filtration

Which one of the following shows the correct order of events?

A 4, 2, 1, 3

B 3, 1, 4, 2

C 4, 1, 2, 3

D 3, 4, 1, 2

Question 7

The receptors for steroid hormones are found:

A inside the target cell because they are hydrophilic

B inside the target cell because they are hydrophobic

C on the plasma membrane of the target cell because they are hydrophilic

D on the plasma membrane of the target cell because they are hydrophobic.

Question 8

Select the statement that best describes an adaptation of plants. Plants have:

A sunken stomata to allow them to stay open on hot, windy days

B parenchyma in the cortex to control the entry of water into the xylem

C vascular tissue to store the products of photosynthesis

D a waxy cuticle to deter mammalian herbivores.

9780170411660

Question 9

A girl contracted influenza during the winter months. The next year in winter, she came in contact with the influenza virus, but this time she did not show any symptoms of the disease. This is because the girl:

A was injected with influenza antibodies when she first contracted influenza

B produced and stored memory B cells specific to the influenza strain of that year

C produced and stored memory B cells that could respond to any strain of influenza

D produced increased numbers of cytotoxic T cells, which destroyed the virus particles.

Question 10

In 2009, the Australian Government authorised a mass immunisation program against swine flu. It was intended that the immunisations:

A would compensate for the quarantine efforts, which failed

B killed the flu particles in which it came into contact

C reduced the virulence of the swine flu

D would prevent the multiplication of the virus in the people who were immunised.

SHORT-ANSWER QUESTIONS

Question 1

a Identify the structures in the following diagram of a typical plant cell.

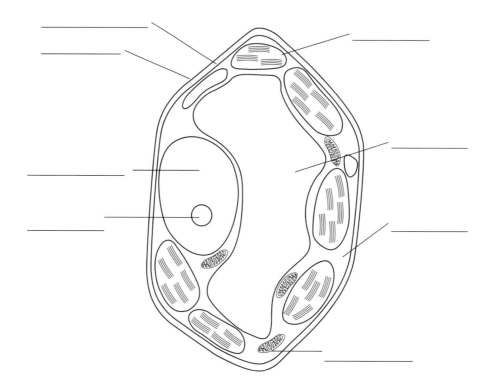

b Name two features that could be used to identify this cell as a:

i eukaryotic cell

_____

ii plant cell.

_____

c Seaweeds can take up iodide ions so vigorously that this the iodide ion is more than a million times more concentrated inside the cells than in the surrounding sea water. Identify the cell structure involved and explain how this difference in concentration is maintained.

_____

_____

_____

d Photosynthesis is divided into two distinct stages. Complete the following table about these stages.

Name of stage	Site within a chloroplast

e Suggest two ways of improving the rate of photosynthesis in some tomato plants that are grown in a greenhouse.

_____

_____

Question 2

The diagrams below show two types of vascular tissue in plants.

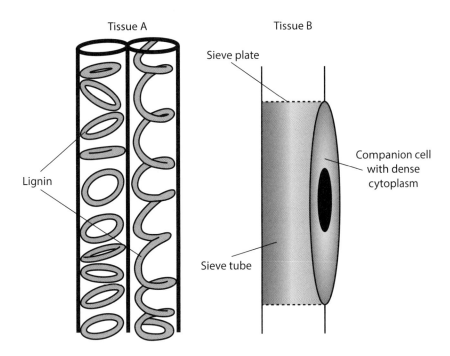

a Name which of the tissues transports water from the roots to the rest of the plant.

_____

b Name one of the factors which causes water to move from the roots to the leaves.

_____

c Describe the main function of the other tissue (that you did not choose in 2a).

_____

Question 3

Antidiuretic hormone (ADH) is a protein that aids in the control of fluid levels in mammals. Osmoreceptors in the hypothalamus detect the variations in the concentration of blood. The hypothalamus causes the release of ADH from the pituitary gland. ADH is carried in the blood to three target tissues: the distal tubules and collecting ducts of the kidneys, the sweat glands, and the smooth muscles of small blood vessels.

Some of the effects of differing levels of ADH on body functions are shown below.

Concentration of ADH (pg/mL)	Output of urine (L/day)	Sweat gland activity	Blood pressure
0.5	15.0	High	Decreased
3.6	1.5	Moderate	No change
4.7	0.5	Low	Increased

a Choose which concentration of ADH is most likely found in a desert mammal and explain how ADH blood concentration affects survival in dry environments.

_____

_____

b Describe and explain the effect on ADH levels in the blood after drinking lots of water.

_____

_____

c Draw a stimulus–response model for water balance in mammals.

d The nervous and endocrine systems work together to maintain homeostasis. Use the following table to describe two ways in which these systems are different.

Nervous system	Endocrine system

Question 4

Strangles is an infectious disease in horses caused by the bacterium *Streptococcus equi*. The bacteria enter the mucous membranes of the upper respiratory tract. Outbreaks may occur when large numbers of horses are gathered together.

a Suggest how the bacteria is spread.

_____

b Identify three other ways disease pathogens are transmitted.

_____

_____

_____

c A vaccine containing dead *Streptococcus equi* is available. To protect against infection, injections are given two weeks apart and followed up with boosters every 12 months. Suggest how the vaccine protects against infection and why more than one injection is given.

_____

_____

_____

d Antibodies to *Streptococcus equi* can be given immediately to reduce infection in a horse but *Streptococcus equi* vaccine is more effective long term. Explain why.

_____

_____

e One of the easiest ways for a pathogen to gain entry into the body is via the gastrointestinal tract, often attached to food. Bacteria such as *Salmonella* cause these foodborne illnesses, commonly referred to as food poisoning. Devise a plan to prevent the spread of *Salmonella*.

_____

_____

CHAPTER 1 REVISION

■ **1.1 STRUCTURE AND FUNCTION OF CELL MEMBRANES**

1

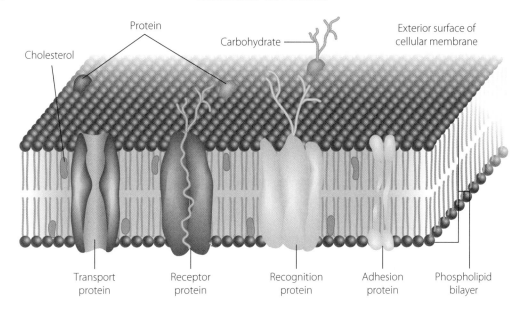

Extracellular environment

Protein

Carbohydrate

Exterior surface of cellular membrane

Cholesterol

Transport protein

Receptor protein

Recognition protein

Adhesion protein

Phospholipid bilayer

Intracellular environment

2 Fluid: phospholipid bilayer allows the membrane to flow and change shape like a two-dimensional fluid.

Mosaic: protein molecules embedded in the bilayer look like a mosaic.

Model: a simplification of a complex natural structure in order to clarify and describe it.

3 Hydrophilic: a substance that tends to interact with and dissolve in water. Hydrophobic avoiding association with water. Phospholipids are molecules that possess both a hydrophilic and a hydrophobic end. The hydrophilic heads turn to point either into the internal or external water-based environments of the cell, while the hydrophobic tails point towards each other, forming the bilayer.

4 Support: technological improvements in visualising cells like the development of the light microscope and the election microscope, as well as ways to calculate the proportion of difference substances in the membrane were vital to the development of the fluid mosaic model.

5

STRUCTURE	FUNCTION
Receptor molecule	Protein that binds hormones and other signal molecules
Carrier molecule	Membrane protein that assists other molecules to cross the membrane in facilitated and active transport
Protein channel	Protein that forms a passageway across the membrane to allow the passage of hydrophobic substances
Recognition molecule	Glycoprotein marker that allows the immune system to distinguish between the body's own 'self' cells, and foreign invaders ('non-self' cells)
Adhesion molecule	Membrane glycoprotein that helps link cells together

■ **1.2 CURING CANCER: CHANGES TO THE CELL MEMBRANE MARK CANCER CELLS FOR DEATH**

1 Phagocytosis: a form of endocytosis in which a dying cancer cell is moved across a membrane by bulk transport. The white cell sends out projections of its membrane that surround the cancer cell and when these membranes meet and fuse, a vesicle is formed that stores and transports the cancer cell in the cytoplasm.

2 Active movement uses energy to move substances across membranes and up a concentration gradient, e.g. phagocytosis. In passive movement, substances move down a concentration gradient without the expenditure of energy.

3 Either holes could form in the phospholipid bilayer to make it leaky, or membrane proteins could be damaged, leaving gaps in the membrane.

4 Membrane proteins.

5 Complementary shapes are like a lock and key, where the shapes match each other, allowing the two molecules to bind to each other.

1.3 OSMOSIS

1 a Isotonic: a liquid with an equal solute concentration to another liquid

b Hypertonic: a solution with a higher solute concentration than another solution

c Hypotonic: a solution with a lower solute concentration than another solution

2 a Very salty water (hypertonic) in the drip means a low concentration of water outside the blood cells. Water moves from a high concentration of water to a lower concentration of water, causing it to diffuse out of the blood cells and making them shrink and become crenated.

b Soaking a blood-stained piece of clothing in cold water is placing the blood cells in a hypotonic solution. Water will diffuse from this high concentration of water into the cells where the concentration is lower, causing the cells to burst, which will enable them to be more easily removed from the clothing.

c Plant tissue, like lettuce, has a cell wall outside its cell membrane. This prevents the membrane from swelling too much when water enters the cell by osmosis. Animal cells have no cell wall; therefore, the cell will swell until it bursts.

1.4 SURFACE-AREA-TO-VOLUME RATIO

1 a $SA:V = \dfrac{150}{125} = 1.2$

b $SA:V = \dfrac{0.54}{0.027} = 20$

2 a

CUBE (mm)	SA (mm²)	V (mm³)	SA:V
0.5	1.5	0.125	12
1	6	1	6
2	24	8	3
4	96	64	1.5

b The ratio becomes smaller as the size of the cube increases.

c The uptake of important materials such as glucose and oxygen into a cell occurs via its cell membrane. These materials are then used to fuel the chemical reactions that occur throughout the volume of the cytoplasm. For a cell to be able to supply the volume of its cytoplasm with its metabolic requirements and remove wastes, it needs a large surface-area-to-volume ratio. The large surface-area-to-volume ratio of smaller cells allows them to obtain their requirements more efficiently than a large cell.

3 a

RECTANGLE	LENGTH (mm)	WIDTH (mm)	HEIGHT (mm)	SA (mm²)	V (mm³)	SA:V
1	32	16	1	1120	512	2.1875
2	16	16	2	640	512	1.25
3	8	16	4	448	512	0.875
4	8	8	8	384	512	0.75

b Surface area decreased and volume is unchanged.

c SA:V decreased as the height increased.

d $32 \times 16 \times 1$

e Leaves are long and flat to provide the greatest SA:V ratio for the uptake of gases (carbon dioxide) essential for photosynthesis.

1.5 SECOND-HAND DATA ANALYSIS: OSMOSIS IN POTATOES

1 Independent variable: potato boiled or not boiled; control: potato well without honey

2 Unboiled with honey: water moved by osmosis from the cells to the well from a high concentration of water in the cells to a lower concentration of water in the honey.

No honey: no difference in water concentration so no diffusion of water

Boiled: differentially permeable cell membranes damaged by boiling, so osmosis did not occur

3 The same results would be expected from other types of vegetables, such as carrots, because they too have differentially permeable cell membranes.

CHAPTER 1 EVALUATION

1 A

2 C

3 a Water

  b Sugar

4 a 2% salt: dilute; 5% salt: concentrated

  b Water will diffuse from left to right, from a high concentration of water (2% salt) to a lower concentration of water (5%).

5 Cholesterol is a structural component of membranes that stabilises and strengthens the membrane and maintains it at a suitable fluidity.

6 $SA:V = \dfrac{24}{8} = 3$

7 a Sodium or chloride

  b adenosine triphosphate (ATP)

  c Membrane transport proteins use ATP energy to move the ion from a low concentration (inside cell) to a high concentration (outside cell).

CHAPTER 2 REVISION

■ 2.1 FUNDAMENTALS OF CELL FUNCTION

1 All cells require *energy*. Autotrophic organisms make their own energy-containing food through *photosynthesis*. *Heterotrophs* are organisms that must consume others to gain energy. *Respiration* is the metabolic pathway that breaks down *food* to provide energy to the cell.

2 The initial *eukaryotes* appeared approximately 2 billion years ago. The *endosymbiotic* theory proposes that the first eukaryote cells were created when a cell was *engulfed* by another primitive *prokaryote*. The engulfing of a smaller bacterial cell that survives in the host is known as *endosymbiosis*. It is thought that *chloroplasts* and *mitochondria* were created in the same way, forming a symbiotic relationship with the host cell. Both chloroplasts and *mitochondria* can only arise from chloroplasts and *mitochondria* respectively and both have a *double* membrane and their own *DNA* (genetic material).

3 *Photosynthesis* is the chemical reaction that harvests the Sun's energy and converts it into useable energy. The green pigment *chlorophyll*, found in the *chloroplasts* of plant cells, is able to harvest the Sun's energy and make it available for use in photosynthesis. *Chloroplasts* are oval-shaped membrane-*bound* organelles. Photosynthesis is a series of chemical reactions that produces the energy-rich molecule *glucose* and the gas *oxygen*.

4 *Cellular respiration* is the chemical reaction whereby *glucose* is broken down into *carbon dioxide* and *water*, providing energy to the cell. This reaction occurs in organelles called the *mitochondria*. Like *chloroplasts*, mitochondria have a double membrane and their own *DNA* (genetic material).

■ 2.2 ORGANELLES AND MACROMOLECULES IN CELLS

1 a Lysosomes

  b Mitochondria (singular: mitochondrion)

  c Rough endoplasmic reticulum

  d Chromoplast

  e Nucleus

  f Smooth endoplasmic reticulum

  g Leucoplast

  h Plastids

2 Eukaryotic cells are able to carry out many specialised tasks because their cells are divided into different types of membrane-bound organelles. Membrane-bound organelles can confine important, but harmful, molecules, protecting the rest of the cell from their harmful effects, and can carry out specific functions without interfering with other activities elsewhere in the cell. Cells can also keep membrane-bound organelles localised in particular regions of the cytoplasm where they are needed.

3 Cells with densely packed mitochondria would be using significant amounts of energy. This is because the function of mitochondria is to carry out respiration and release energy from food in the form of ATP. They could be muscle cells or sperm cells, or could be carrying out active transport.

4 A monomer is a small molecule that acts as a building block for macromolecules, while a polymer is a large molecule built up by linking monomers together.

5

SUBSTANCE	ROLE
Urea	A nitrogenous waste produced when amino acids are broken down in mammals
Starch	An important energy-storing polysaccharide in plants
Uric acid	A nitrogenous waste product produced by desert animals
Cellulose	A polysaccharide made of glucose subunits that is the main component of plant cell walls
Glycogen	An important energy-storing polysaccharide in animals
Ammonia	A nitrogenous waste produced by marine fish
Glycerol	A molecule that combines with three fatty acids to form fat
Steroid	A type of lipid produced by the smooth endoplasmic reticulum that forms important hormones

■ 2.3 MICROSCOPES, MAGNIFICATION AND SIZE

1 *Light microscope*: light rays from a lamp pass through a thin specimen then two glass lenses, the *objective* and *ocular* (eyepiece) lenses to our eyes.

Transmission electron microscope: much greater *magnification* and *resolution* as it uses an *electron* beam instead of light to pass through the specimen. It is focussed using *electromagnets* instead of glass lenses.

Scanning electron microscope: solid specimens are bombarded with a beam of electrons, providing a lower *resolution*, three dimensional *surface* view of the specimen.

2 B

3 a $100\,\mu m$ and $0.1\,mm$

b Red blood cell ($10\,\mu m$ or $100\,nm$) is 100 times bigger than a virus ($100\,nm$).

c 10 000

d It is valid to say that many objects can be viewed under both the electron microscope and the light microscope, because the diagram shows their ranges to be: light microscope, 1 mm to 100 nm, and electron microscope 10 000 nm to 0.1 nm, which is a significant overlap.

4 a False

b True

c True

d False

■ 2.4 USING THE LIGHT MICROSCOPE TO MEASURE CELLS

1 *Total magnification*: is calculated by multiplying together the magnifications of the objective and ocular (eyepiece) lenses.

Oil immersion lens: a drop of *oil* between a ×100 objective lens and the slide increases the *magnification* and *resolution* possible when viewing very small cells such as bacteria.

The field diameter: is the *diameter* of the field of view, the circle of light, seen down the microscope. A *ruler* or mini-grid can be used to measure the *field* of view of a microscope at low power. As magnification *increases*, the field of view *decreases*. To find the field diameters at higher magnifications, divide the *field* diameter at low power by the change in total *magnification*.

Cell size: If the diameter of the field of view is known, the size of a cell can be estimated as follows:

$$\text{Cell size} = \frac{\textit{diameter} \text{ of field of view}}{\textit{number} \text{ of cells across field}}$$

2

| POWER | LENS COMBINATION | | TOTAL MAGNIFICATION | FIELD DIAMETERS | |
	OCULAR LENS	OBJECTIVE LENS		MILLIMETRES (mm)	MICROMETRES (μm)
Low power	× 10	× 10	× 100	1.25	1250
High power	× 5	× 40	× 200	0.625	625
Oil immersion	× 10	× 100	× 1000	0.125	125

3 a 0.25 mm or 250 μm

   b Half of one of these cells would be seen with the ×10 ocular and oil immersion lens.

CHAPTER 2 EVALUATION

1 A

2 B

3 D

4 0.0086 mm

5 Any two from similarities and any two from differences.

COMPARISON	FEATURE	PROKARYOTES	EUKARYOTES
SIMILARITIES	Outer boundary	Phospholipid membrane	Phospholipid membrane
	Ribosomes	Present	Present
	Genetic material	DNA	DNA
	Site of chemical reactions	Cytosol	Cytosol
DIFFERENCES	Size	Length: 1–10 μm	Length: 10-100 μm
	Chromosomes	Single, circular	Multiple, linear
	Unicellular/multicellular	Only unicellular	Unicellular and multicellular
	Nucleus	Absent	Present
	Date observed in fossil record	3.5 billion years ago	1.8 billion years ago
	Cellular organisation	Cytoplasm not compartmentalised	Compartments formed by membrane-bound organelles

6 Lysosomes contain digestive enzymes that split complex chemical compounds into simpler ones. They also remove and digest old worn out organelles.

7 Endosymbiosis

8 Heterotrophs cannot synthesise their own organic compounds from simple inorganic material. Instead, they depend on other organisms for nutrients and energy requirements. Autotrophs are able to produce their own food from light energy through photosynthesis.

CHAPTER 3 REVISION

3.1 INTERNAL MEMBRANES IN CELLS

1 If the box used is 5.2 cm × 3.5 cm × 1.5 cm, the total surface area is 20.4 cm².

A sheet of A4 paper is 21 cm × 29.7 cm. The total surface area is 623.7 cm².

The A4 sheet of paper is 30.5 times bigger than the box.

2

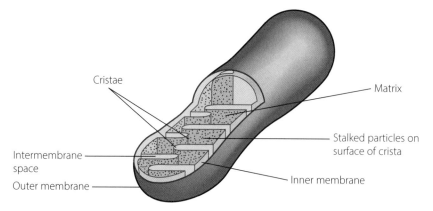

3 Similarities include an outer layer with an inner folded structure; a large surface area; compartmentalisation of areas.

Differences include shape – sausage shape in sketch, rectangular shape in model; inner membrane with stalk-like structures in sketch, continuous folding of inner layer in model; matrix made of protein-rich fluid in sketch, air in spaces in model.

4 Without internal compartmentalisation, the matches move around and mix with each other. This would be the same for chemical reactions. There would be interference between reactions in the same conditions. Once the internal folded paper is added, the matches don't touch each other so there can be no mixing and there could be different conditions in each fold. Due to the structure of mitochondria, chemical reactions are isolated from each other and can occur at the same time in different conditions.

5 Students reflect on how useful this activity was for their personal understanding of the concept. Generally models are useful in science to represent something too large or too small to be seen; to explain something complex in a simple manner; to make predictions of expected results.

■ **3.2 ENZYME ACTION**

1 a Metabolism: the sum of all chemical reactions within the organism

b Catalyst: a substance that is used to speed up a chemical reaction without being used up in the reaction

c Activation energy: the energy required to initiate a reaction

2

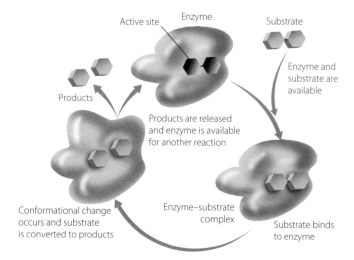

3 The induced fit model explains that the active site of an enzyme can undergo specific changes induced by the substrate to achieve a high degree of specificity with the substrate. As the bonds of the substrate are stretched by molecular interactions, the energy required to kick-start the reaction is significantly lowered.

4 a Lysozyme: pH 4–5. Pepsin: pH 2. Trypsin: pH 6–7

b The small intestine is alkaline (basic). According to the graph, pepsin stops functioning effectively outside an approximate pH range of 3.5–4. Therefore, as pepsin enters into the small intestine the pH changes and causes the enzyme to denature. Due to the active site being changed the enzyme can no longer function.

5 A non-competitive inhibitor is a molecule that binds to an enzyme at site other than the active site. It changes the shape of the enzyme, which prevents the substrate from binding at the active site. When there is a high concentration of product in the cell, the non-competitive inhibitor will bind with the enzyme to stop the synthesis of the product. When the products are removed from the cell, the inhibition will decrease and the enzyme will move back into the active state. The non-competitive inhibitor helps to regulate the concentration of products in the cell.

■ 3.3 PROPERTIES OF ENZYMES

1 When an enzyme is folded into its tertiary structure, a specific shape is created on its surface. This particular shape forms the active site.

2 a Starch

   b Glucose

   c

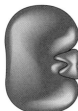

3 If the enzyme's shape changes so much to break hydrogen bonds between connecting units of amino acids, proteins cannot return to their original shape, causing them to become denatured.

4 If a patient's temperature increases, enzymes change shape. If they change to such an extent that they are no longer able to function, important reactions in the patient's body may not proceed at a fast enough rate to maintain life.

5 a i Product A

   ii Products B, C and D

   b Product B would be in excess; products C and D would be missing.

   c An enzyme or cofactor acting on the initial reactants might be missing or there might be a competitive inhibitor of the enzyme acting on the initial reactants.

   d If any of the enzymes were faulty, D would not be produced.

   e Levels of product B would be tested to see if it was in excess and levels of product C would be tested to see if it was missing.

6 Enzymes:

- generally work very rapidly
- are not destroyed or altered by the reactions that they speed up, so they can be used again
- can speed up reactions in either direction
- are usually specific to particular reactions.

■ 3.4 FACTORS AFFECTING ENZYME ACTION

1 Enzymes work efficiently within a narrow range of pH. If the pH varies, enzyme function is restricted. Therefore, it is important to keep the surroundings of cells at an optimum pH for enzyme function.

2 a The amount of enzyme is the factor limiting the rate of the reaction because there are no more active sites available.

   b Doubling the amount of enzyme would double the rate of reaction before levelling off.

3

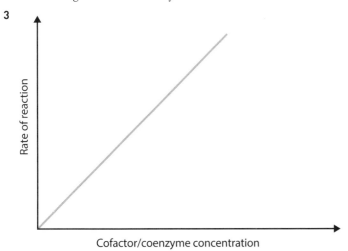

4 If these types of heavy metals are in your body, they will bind to the active sites of enzymes. This stops substrates binding and over time, chemical reactions slow down to a dangerous level.

1 **a** Hydrogen peroxide (H_2O_2)

 b Catalase

 c Water (H_2O) and oxygen (O_2)

2 Active site

3 Compartmentalising a cell by having membrane-bound organelles creates specialised environments for specific functions. This enables a large number of activities to occur at the same time in a very limited space and under different conditions.

4 pH has a profound effect on the shape of the activity site and hence the enzyme's activity.

5 **a** The body temperature of humans is 37°C.

 b Test tube at 70°C: this high temperature is likely to denature the enzyme. Starch would not be broken down into maltose.

 Test tube at 10°C: this low temperature is likely to inactivate the enzyme. Starch would not be broken down into maltose.

 c After returning to 37°C, amylase in the test tube exposed to 10°C would become active again. Starch would be broken down to maltose. Amylase in the test tube exposed to 70°C has been denatured which is a permanent change in the active site. Starch would not be broken down to maltose.

 d pH, substrate and enzyme concentration, inhibitors and cofactors and coenzymes all affect the rate of enzyme-catalysed reactions.

CHAPTER 4 REVISION

■ **4.1 ATP−ADP CYCLE**

1

ATP molecule   ADP molecule

2

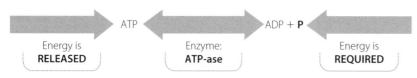

3 ATP

4 Energy is stored in the chemical bond.

5 When a cell requires energy to drive a reaction, the high energy chemical bonds attaching the last phosphate group to ATP are broken, thus releasing stored energy. This energy is now available to fuel a cellular reaction. Free energy obtained from a cellular reaction can be used to add a phosphate group to ADP, converting it to ATP.

6 The ATP−ADP cycle allows the cell to shuttle energy between reactions. It provides the cell with an efficient linking or coupling of energy-yielding processes to energy-requiring processes within the cell by conserving, transferring and releasing energy.

7

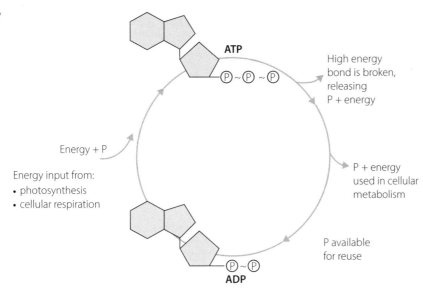

■ 4.2 CHEMICAL REACTIONS: PHOTOSYNTHESIS

1 Light-dependent stage: chlorophylls, carotenoids and xanthophylls absorb light energy within the thylakoid membranes. Electrons within the pigments become energised. This energy is used to split water molecules to form hydrogen ions and oxygen. The oxygen is a by-product of the reaction and is released at this stage. ATP is formed in this stage by the bonding of ADP and P.

Light-independent stage: the light-independent reactions occur in the stroma, the gel-like matrix within the chloroplast. Hydrogen ions and ATP formed in the light-dependent stage are used to convert carbon dioxide, taken in from the atmosphere, into glucose. When ATP is used it is broken down into ADP and P, which can be reused in the light-dependent stage. The glucose is the monosaccharide used to build complex polysaccharides such as cellulose and starch.

2 a The chloroplasts were suspended in the water solution. Some of the water molecules included labelled oxygen molecules. The chloroplasts used the water molecules from the solution in the light-dependent stage. When the water molecules were broken down in the reaction the labelled oxygen was released into the airspace.

   b The solution is expected to become colourless. When the water molecule is split in the thylakoid, oxygen is given off and hydrogen ions are formed. The hydrogen ions then reduce the hydrogen acceptor and as a result the solution becomes clear.

■ 4.3 CHEMICAL REACTIONS: AEROBIC CELLULAR RESPIRATION

1 glucose + oxygen → carbon dioxide + water + energy

2 $C_6H_{12}O_6 + 6O_2 \rightarrow 6CO_2 + 6H_2O + 36–38ATP$

3

STAGE OF AEROBIC RESPIRATION	INPUT	OUTPUT	ATP MOLECULES FORMED FROM ONE GLUCOSE MOLECULE	LOCATION IN CELL
Glycolysis	Glucose	Pyruvate	2	Cytosol
Kreb's cycle	Pyruvate	Carbon dioxide	2	Mitochondrial matrix
Electron chain transfer	Oxygen	Water	32–34	Mitochondrial cristae

4

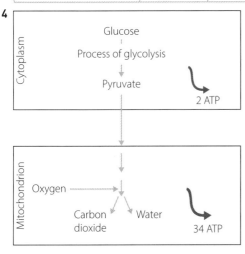

■ **4.4 CHEMICAL REACTIONS: ANAEROBIC CELLULAR RESPIRATION**

1 glucose → ethanol + carbon dioxide + adenosine triphosphate

2 glucose → lactic acid + adenosine triphosphate

3 At the start of the run, the blood would be supplying his cells with oxygen. Cellular respiration would use the oxygen to break down glucose and release the 36 molecules of ATP. As the man continued through the training run, the exercise became more strenuous. The body could not deliver the required amount of oxygen to the cells, and continue providing the cell with energy, so the body started to undergo anaerobic respiration, also known as lactic acid fermentation. After the man finished the training run, oxygen would once again become available; the lactic acid would then revert back to pyruvate for aerobic respiration.

4 Students' answers may vary. Students could make the following arguments:

Correct: most of the reactions for the biochemical pathway occur in the mitochondria, producing the largest portion of energy; therefore, it can be considered as the site of cellular respiration.

Incorrect: glycolysis, the starting reaction of cellular respiration, occurs in the cytoplasm to form pyruvate and two molecules of ATP. The pyruvate then moves into the mitochondria where it reacts with oxygen to form 34 ATP molecules, carbon dioxide and water. While most of the reactions occur in the mitochondria, not all of them do; therefore, the mitochondria is the main site for cellular respiration but not the only site required for the pathway.

5 Aerobic cellular respiration makes available a lot more energy (36–38 ATP) compared with anaerobic cellular respiration (2 ATP).

6

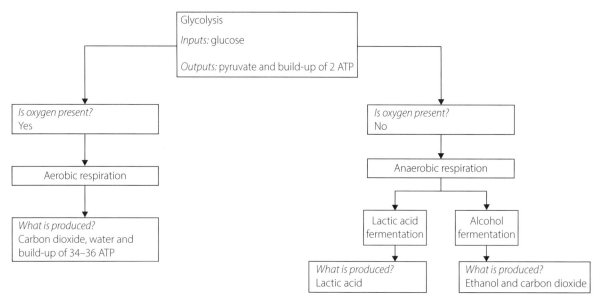

■ **4.5 PHOTOSYNTHESIS AND CELLULAR RESPIRATION**

1

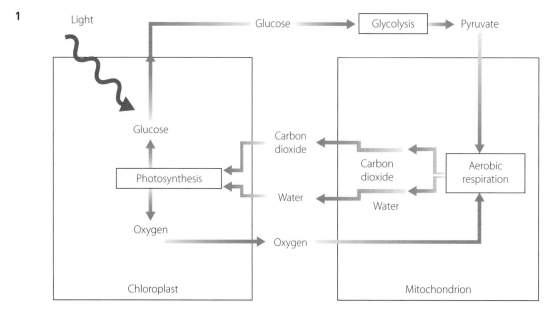

2 a

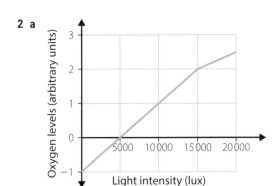

b O_2 is an output of photosynthesis. O_2 is an input of aerobic cellular respiration. As the light intensity increases the rate of photosynthesis increases and oxygen levels rise. In the absence of light photosynthesis ceases. Aerobic cellular respiration occurs at a constant rate regardless of the light intensity.

c

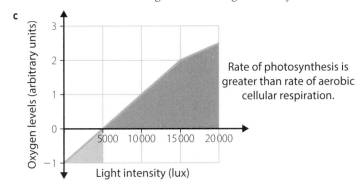

Rate of photosynthesis is greater than rate of aerobic cellular respiration.

Rate of photosynthesis is less than rate of aerobic cellular respiration.

d Rate of photosynthesis = rate of aerobic cellular respiration

e Oxygen is continually used in cellular respiration but is produced in photosynthesis. The higher the oxygen concentration, the greater the rate of photosynthesis.

CHAPTER 4 EVALUATION

1 A

2 Glycolysis occurs in the cytoplasm of all plant and animal cells.

3 Cellular respiration occurs throughout a 24-hour day.

4 a Light-independent stage

   b Light-dependent stage

5 The product of fermentation in bacteria and yeast is ethanol and carbon dioxide. The product of fermentation in an animal cell is lactic acid. Yeast is used for bread making because the carbon dioxide gas produced in fermentation bubbles through the dough, causing it to rise. In lactic fermentation, there is no liberation of carbon dioxide; therefore, the desired result cannot be obtained.

6 ATP couples energy-releasing reactions with energy-requiring ones.

7 The experimental design will depend on students' answers. It should include exposing both tubes to the same conditions and measuring the amount of growth. The tube with yeast cells only capable of performing aerobic respiration will grow more than the tube with yeast cells that can carry out anaerobic cellular respiration. This is predicted because aerobic respiration releases much more energy than anaerobic respiration. If the tubes are deprived of oxygen, the reverse result will be expected. This is because aerobic respiration will not occur if oxygen is absent.

8 a The algae will not be able to photosynthesise and will eventually die.

   b The jellyfish will not be accessing the products of photosynthesis from the algae but will still be using food from other sources. Their supplies of energy will diminish but they will still survive.

CHAPTER 5 REVISION

■ **5.1 PROPERTIES OF STEM CELLS**

a Unspecialised

b Replicating

c Differentiate, zygote, totipotent, embryo, adult

■ **5.2 STEM CELL DIFFERENTIATION**

1 Only a small fraction of all the genetic material within each cell is activated or 'switched on', so that only particular proteins are produced as a result to bring about particular structural and functional characteristics in that cell type.

2 For example, for a neuron (nerve) cell in brain:

- oxygen must be delivered to the neuron by red blood cells for cellular respiration reactions to provide energy
- heart cells must contract and relax to pump red blood cells through blood vessels made up of cells
- oxygen enters blood by diffusing across cells lining the inside of lungs – cells in 'breathing muscles' contract and relax for inhalation and exhalation of air into/out of lungs
- glucose is also necessary and it passes through cells that make up the wall of the small intestine into capillaries
- glucose is produced by the action of digestive enzymes manufactured by pancreatic cells breaking down carbohydrates.

■ **5.3 STRUCTURAL HIERARCHY IN MULTICELLULAR ORGANISMS**

1

1 no organelles

2 eukaryote

3 bacteria

4 completely functional single cell

5 multicellular

6 *Volvox*

7 group of specialised cells working together to perform particular function/s

8 organ

9 heart

10 collection of organs working together to perform particular function/s

2 cells → tissues → organs → systems

CHAPTER 5 EVALUATION

1 D

2 C

3 Stem

4 Divide or replicate, differentiated, functions,

5 Different specialised cells, initially from as far back as in the developing embryo, have different segments of their genetic material or chromosomes; that is, different genes, either 'switched on' to manufacture particular proteins to perform particular functions, or 'switched off' if that particular cell does not carry out that particular function.

CHAPTER 6 REVISION

■ **6.1 STRUCTURE AND FUNCTION OF GASEOUS SURFACES**

1 Colour 'from pulmonary artery' arrow *blue* because arteries carry blood away from heart, but pulmonary artery leaves heart to carry *deoxygenated* blood to lungs, where it becomes oxygenated. This blood leaving heart, has previously travelled around the body, where oxygen has diffused *out* of capillaries to all body cells.

Colour 'to pulmonary vein' arrow *red* because veins return blood to heart and pulmonary vein returns newly oxygenated blood from lungs to heart to be distributed to all cells.

2 The pulmonary artery is the only artery carrying deoxygenated blood.

9780170411660

3 Red blood cells nearest 'From pulmonary artery' arrow: *blue* (fourth cell from arrow perhaps starting to turn a little 'purplish' as O_2 diffuses into capillary from alveolus, and CO_2 continues to diffuse out).

Red blood cells nearest 'To pulmonary vein' arrow: *red*, holding highest O_2 concentration (fourth cell from arrow perhaps still being 'purplish' as CO_2 concentration in capillary continues to fall and O_2 concentration continues to increase along capillary).

4 O_2: diffuses from alveolus into capillary; CO_2: diffuses from capillary into alveolus

5 Mouth/nose → throat → voice box → trachea → bronchus → bronchiole → alveolus

6

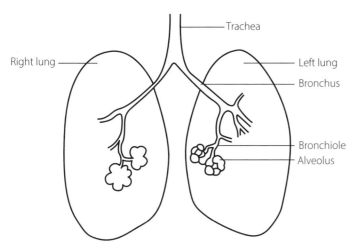

7 Similarities:
- alveoli walls and covering of gill only one cell thick
- very large surface area of multiple 'bunch of grape-like' alveoli, and of layers of filaments of gills
- outer surface of alveolus bathed in extracellular fluid, inner surface of alveolus lined with film of fluid
- outside surface of gill bathed in surrounding water of external environment, inner surface lined with fluid
- capillaries forming network surrounding alveoli are only thousandths of millimetre distance away; capillary network fills interior of gills – both: capillaries very close.

8 Differences:
- alveolus is inside the lung within the body to prevent dehydration
- gill is on the exterior of body bathed in the surrounding external environmental water.

9 External gills suit fish to aquatic environment where drying of exchange surface is *not* an issue, as it is for gas exchange surface in terrestrial mammal

■ 6.2 STRUCTURE AND FUNCTION OF CAPILLARIES

1

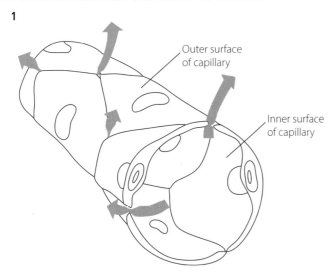

2

CAPILLARY STRUCTURE	FUNCTION
Smooth inner surface	Minimal frictional resistance to blood flow
Wall only one cell thick	Easily allows diffusion
Extensive branching	To reach within thousandths of millimetre of every body cell
Very small diameter	To reach within thousandths of millimetre of every body cell

CHAPTER 6 EVALUATION

1 D

2 A

3 Counter-current

4 Haemoglobin

5 a CO_2, O_2, digested food products, cellular waste products, hormones

   b **i** Diffusion: movement of molecules, without need for energy, from area where they are in high concentration to area where they are in lower concentration.

      ii Concentration gradient: difference in concentrations of particles in one area, or on one side of a membrane, compared to that in another area or on other side of membrane.

CHAPTER 7 REVISION

■ **7.1 ABSORPTION OF NUTRIENTS**

1 and **2**

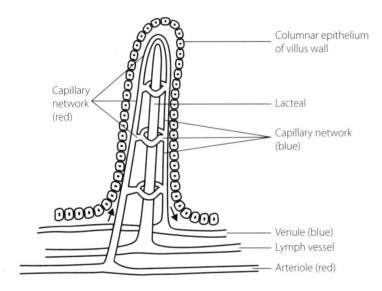

9780170411660

7.2 DIGESTION OF FOOD

	ORGAN	IS ORGAN A SITE OF CHEMICAL DIGESTION? (YES OR NO)	IF YES	
			TYPE OF ORGANIC MOLECULE	ENZYME TYPE
	Mouth	Yes	Carbohydrate	Carbohydrase/amylase
	Oesophagus	No		
	Stomach	Yes	Protein	Protease (pepsin)
Small intestine	Duodenum	Yes	Protein	Proteases – pancreatic (trypsin, peptidase)
			Carbohydrate	Amylase – pancreatic
			Lipid	Lipase – pancreatic
	Jejunum	Yes	Protein	Protease – intestinal peptidase
	Ileum	No		

7.3 NITROGENOUS WASTES

Ammonia: B, C, F

Urea: E, G

Uric acid: A, D, G

7.4 NEPHRONS

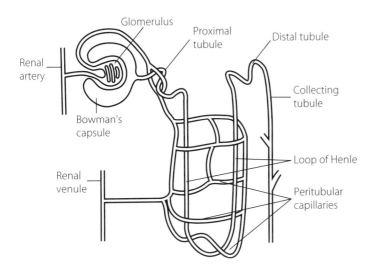

7.5 PRODUCTION OF URINE

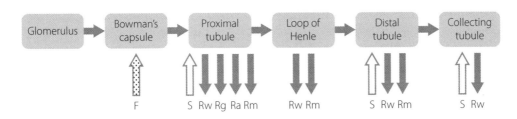

CHAPTER 7 EVALUATION

1 B

2 C

3 a Alveolus: contains air that is external environment

Villus: ingested food in small intestine, of which villus is wall, is external to the body

Nephron: interior of Bowman's capsule and all following tubules: part of external environment, as they connect directly with ureter, bladder and urethra, out of which wastes are excreted from body

Glomerulus and peritubular capillaries (surrounding nephron tubules): part of internal environment

b Alveolus: alveolar wall then capillary wall

Villus: wall of villus then wall of capillary or lacteal

Nephron: glomerular wall then Bowman's capsule wall

c • Only two, one-cell-thick surfaces for molecules to cross

• Exchange surfaces moist

• Increased surface areas for exchange

• Capillaries for transport of molecules in very close proximity

d • Surface area of alveolar and villi walls increased in three dimensional manner; surface area of nephron tubules increased lengthwise, though small intestinal surface area, of which villi form lining, increased longitudinally also

• movement of molecules in - alveoli and villi by passive diffusion; nephron by both passive and active transport; nephron from internal to external environment.

CHAPTER 8 REVISION

■ 8.1 STOMATA

1

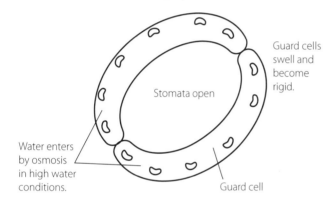

Turgid position

Guard cells swell and become rigid.

Stomata open

Guard cell

Water enters by osmosis in high water conditions.

2

Flaccid position

Water leaves by osmosis in low water conditions

Stomata close as guard cells lose water and soften

Guard cell

3 Guard cells and the stomata they operate provide the entry and exit points for gases in leaves. When water conditions are good, the stomata remain open for gases to diffuse freely into and out of the leaf, including water vapour. As water levels drop, the guard cells become softer and the stomata close, cutting off the leaf's access to gases for photosynthesis, but also protecting it against further water vapour loss.

4

STRENGTHS	WEAKNESSES
Air in the balloon is a good analogue for water in the guard cell – more air results in a firmer balloon with an opening, while losing air causes it to deflate and close up.	One balloon is trying to represent two cells. Water enters and leaves guard cells by osmosis rather than by force.
It clearly shows the relationship between guard cells and the stoma and emphasises that stomata are not structures - they are spaces between structures.	It is difficult to show the stomata opening, as you need to blow up the balloon while holding it in a circle.

9780170411660

8.2 FACILITATING GAS EXCHANGE

1–4

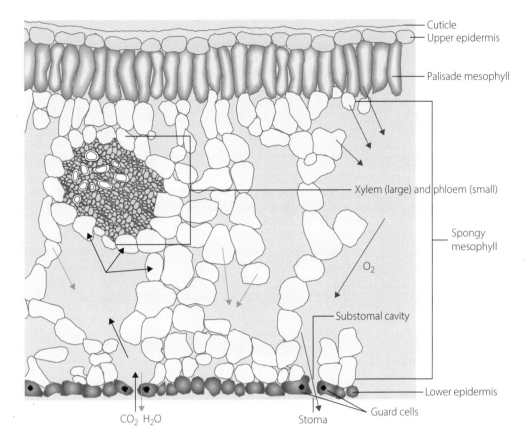

8.3 RELATIONSHIP BETWEEN PHOTOSYNTHESIS AND THE MAIN TISSUES OF LEAVES

1

COMPONENT	STRUCTURE	FUNCTION
Cuticle	A thin layer of transparent wax covering the outer cells of the leaf	To provide an airtight and watertight seal for the cells of the leaf, to protect them from dehydration and overexposure
Epidermis	A thin layer of flat, clear cells on the very outer surface of the leaf, both upper and lower	To separate the cells of the leaf from the outside and to protect the photosynthetic cells, while allowing light through
Palisade mesophyll	A single layer of chloroplast-heavy, columnar cells that are tightly packed beneath the upper epidermis	To conduct the bulk of the photosynthesis for the leaf
Spongy mesophyll	A loosely packed area of irregularly shaped, photosynthetic cells with interconnecting air spaces beneath the palisade mesophyll	To conduct the remainder of the photosynthesis while forming air spaces for gases to diffuse freely throughout the leaf
Phloem	Columns of sieve tube cells with few organelles that are connected to each other with sieve plates to share cytoplasm Companion cells are attached to the sieve tube cells with plasmodesmata	Sieve tube cells carry sugars produced by the mesophyll to other parts of the plant Companion cells provide the bulk of the cellular functions to keep the sieve tube cells alive
Xylem	Columns of empty, dead cells. Tracheids are elongated and overlap each other, while vessel elements are stacked on top of one another	To carry water and dissolved minerals from the roots to the leaves
Stomata	Openings between pairs of guard cells on the underside of the leaf	To regulate the flow of gases into and out of the leaf, including minimising water loss

2 Students' answers will vary but should be reasonable and well argued.

3 Students' answers will vary but should consider the general climate (hot, dry and intensely sunny), which would contribute to fewer stomata and vertically hanging leaves that do not have an 'upper' and 'lower' side, giving a more symmetrical cross-section than normal. A more even distribution of mesophyll shows that photosynthesis is equally important from all sides of the leaf.

8.4 XYLEM AND PHLOEM

FEATURE	XYLEM	PHLOEM
STRUCTURE Describe the cells it is made of and how the cells are arranged together.	Made of tracheids and vessel elements that overlap each other in wonky tubes. Cells are dead and hollow.	Made of sieve tube cells and companion cells. Sieve tube cells are stacked in columns and joined by sieve plates. They are living and share cytoplasm with each other but have few organelles. Companion cells keep the sieve tube cells alive through plasmodesmata connections.
FUNCTION Describe what it transports, in which direction and the forces that drive the movement.	It transports water and dissolved minerals from the roots to the leaves. Root pressure, capillary action and the transpiration stream force the water up the plant.	It transports cytoplasm and all that is dissolved in it, including sugars, from the leaves to other areas of the plant. The concentration gradient forces the sap to flow from areas of high sugar concentration (leaves) to areas of low sugar concentration (high growth areas).
LOCATION Describe its location within the cross-section of a plant stem. (Include both monocot and dicot stems.)	Xylem and phloem are both gathered into vascular bundles. Monocot stem: Vascular bundles are scattered evenly throughout the stem. Dicot stem: Vascular bundles are organised in a single ring near the outer edge of the stem. Xylem oriented in the bundle towards the interior and phloem on the outside.	

8.5 RATE OF TRANSPIRATION

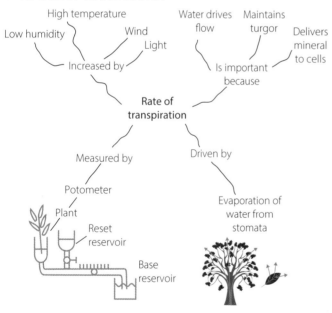

8.6 TERRARIA

Water: Any added water would percolate through to the gravel reservoir from where it would moisten the moss and soil. The water would be taken up by the roots of the plants and pulled up to the leaves by the transpiration stream. After evaporating from the leaves, the water vapour in the air would condense on the lid and sides of the container and drip back down into the soil and moss to be taken up again.

Gases: carbon dioxide used and oxygen produced by the leaves during photosynthesis would skew the concentration of these gases to the oxygen side during the day. During the night, when the leaves are not photosynthesising, the plant is still conducting respiration, which takes in oxygen and produces carbon dioxide. This pushes the concentration of gases back towards carbon dioxide, ready for the next day. Water vapour would be a constant concentration due to the relatively high humidity levels required for the water to cycle appropriately.

1 C

2 A

3 Xylem and phloem

4

Sugars (red)   Water (blue)

5 Photosynthesis requires light to reach the chloroplasts. This means that the epidermis and cuticle must be thin and transparent to allow maximum light through. The palisade mesophyll is dense with chloroplasts to catch all of the sunlight that falls on the upper surface. The cells are cylindrical, with their small ends facing the sun, so that the maximum number of cells can fit on the surface and tunnel the light into the body of the leaf. The spongy mesophyll on the underside of the leaf still contains chloroplasts to take advantage of reflected light from the ground and periods when the leaf is upside down due to wind etc. The air spaces between the cells of the spongy mesophyll allow gases to diffuse freely inside the leaf, which ensures each cell has access to adequate carbon dioxide for photosynthesis.

6 The left jar is transparent, so photosynthesis will still occur. This means the oxygen concentration will increase under the jar and carbon dioxide will decrease. The plant should still be healthy. The jar would act as a greenhouse and trap heat, increasing the rate of transpiration of the plant and thus, the concentration of water vapour will be increased. Water may even condense on the jar surface and run down to the oval surface. The plastic bag over the pot will prevent water from dripping back into it, leading to minor water stress for the plant but one day shouldn't affect the plant's health too badly.

The right jar is painted black, so photosynthesis would not be occurring. The plant would still be respiring, so the concentration of carbon dioxide would increase and the concentration of oxygen would decrease. The black jar would trap heat considerably faster than the transparent jar. This would increase transpiration and water loss by the plant. The water vapour concentration under the jar would be very high. The plastic bag would still prevent water from dripping back into the pot, leading to water stress for the plant. The extreme heat may also kill parts of the plant, if not all of it.

9.1 STIMULUS–RESPONSE MODEL

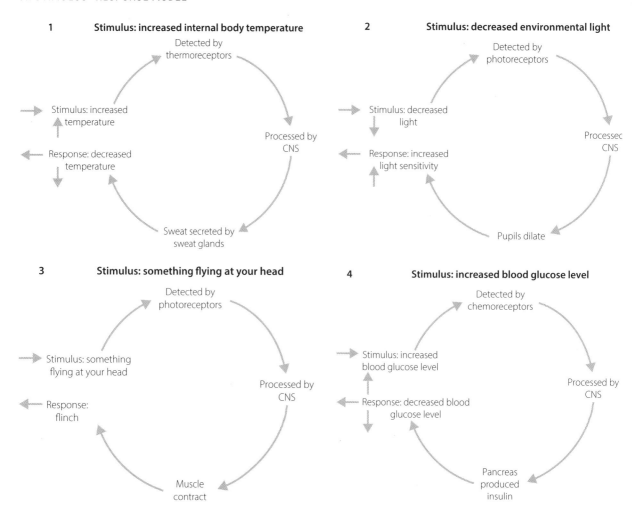

1 Stimulus: increased internal body temperature

Detected by thermoreceptors

Processed by CNS

Sweat secreted by sweat glands

Response: decreased temperature

Stimulus: increased temperature

2 Stimulus: decreased environmental light

Detected by photoreceptors

Processed CNS

Pupils dilate

Response: increased light sensitivity

Stimulus: decreased light

3 Stimulus: something flying at your head

Detected by photoreceptors

Processed by CNS

Muscle contract

Response: flinch

Stimulus: something flying at your head

4 Stimulus: increased blood glucose level

Detected by chemoreceptors

Processed by CNS

Pancreas produced insulin

Response: decreased blood glucose level

Stimulus: increased blood glucose level

9.2 SENSORY RECEPTORS

CATEGORY	LOCATION	FUNCTION
Chemoreceptor	Lining blood vessels and the mouth	Detect the concentration of chemicals in the body
Thermoreceptor	Skin and throughout the body	Detect the temperature of the body
Mechanoreceptor	Mostly in the skin	Detect pressure
Photoreceptor	Retinas	Detect light
Nociceptor	Throughout the body	Detect intense/dangerous levels of temperature, pressure, light or chemicals

9.3 EFFECTORS

1

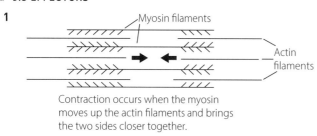

Myosin filaments

Actin filaments

Contraction occurs when the myosin moves up the actin filaments and brings the two sides closer together.

2 The pancreas is an effector in the body because it is a gland that secretes two main substances – insulin and glucagon. These molecules regulate the uptake and release of glucose in the blood. Secretion is controlled by chemoreceptors in the pancreas itself, which detect blood glucose levels and trigger secretion without the influence of the central nervous system (CNS).

3 Growth hormone triggers growth.

Thyroid-stimulating hormone prompts thyroid to secrete other hormones.

Adrenocorticotropic hormone triggers adrenal glands to release cortisol.

Beta-endorphin has analgesic properties.

Prolactin triggers milk production.

Lutenising hormone triggers ovulation or testosterone production.

Follicle-stimulating hormone regulates puberty and reproductive processes.

Melanocyte-stimulating hormone triggers production of melanin.

Antidiuretic hormone prompts kidneys to retain water and constricts blood vessels.

Oxytocin is involved in emotional bonding.

4 The pituitary gland is called the 'master gland' because it secretes many different hormones that play important roles in almost all areas of growth, maintenance and reproduction. Also, many of the hormones it secretes signals other glands in the body to produce their own hormones.

■ 9.4 FEEDBACK CONTROL DIAGRAMS

1 Cardiovascular system increases vasoconstriction; excretory system increases renal absorption of fluid; endocrine system produces vasoconstrictors; digestive system increases fluid reabsorption.

2

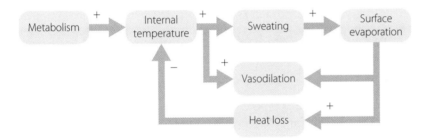

■ 9.5 METABOLISM

1 **a** Metabolism: all of the chemical reactions involved in sustaining life

b Catabolism: biochemical reactions that break complex molecules into simpler ones, releasing energy

c Anabolism: biochemical reactions that build complex molecules from simpler ones, using energy

d Basal metabolic rate: the amount of energy used by the body when completely at rest

e Kilojoules/calories: units of energy, often used in reference to energy obtained through the consumption of food

2 Metabolism is all of the chemical reactions that sustain life. If someone's metabolism was unusually fast, that would mean their reactions proceeded unusually fast. This would mean that the energy the food they eat would be released quickly and used quickly. However, it doesn't mean that the energy they require to run their bodies is more or less than any other person. If someone did have faster reactions, they would need to eat less, because the energy would be released more efficiently. If someone actually had a fast metabolism, they would need to be more careful with what they ate, which is the opposite to the general usage of the term.

3 Students' answers will vary but could include optimising body pH by deep breathing, improving enzyme access to substrates through improving blood flow by regular cardio exercise and well-fitting clothing and maintaining appropriate body temperature through smart environment choices.

CHAPTER 9 EVALUATION

1 C

2 D

3 Catabolism could be any of the digestive reactions including glycolysis, while anabolism could be any of the structure building reactions including gluconeogenesis.

4

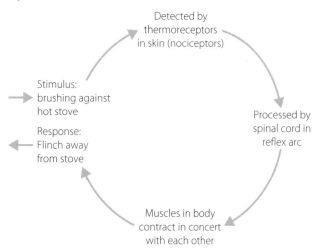

Stimulus: brushing against hot stove

Detected by thermoreceptors in skin (nociceptors)

Processed by spinal cord in reflex arc

Muscles in body contract in concert with each other

Response: Flinch away from stove

5 Receptors detect stimuli and monitor the environment, both internal and external. Homeostasis seeks to maintain body conditions within narrow limits and the body cannot do this without monitoring the conditions. Therefore, receptors are vital to providing information to the body about what is happening, so that the body can respond appropriately.

6 Positive feedback is a cyclic process where responses reinforce and strengthen disturbances to normal cellular function. This reinforcement of a disturbance increases the disturbance, rather than counteracting it and returning the system to normal limits. This is completely opposite to the goal of homeostasis, which is to keep conditions as close to static as possible. Positive feedback is necessary for development, by disrupting homeostasis so the body can find a new normal in the next phase of development, but this is still a disruption of homeostasis.

CHAPTER 10 REVISION

■ 10.1 CELLS THAT TRANSPORT NERVE IMPULSES

1 a Axon: the tubular extension of a neuron cell body that conducts the nerve impulse

  b Soma: the main cell body of a neuron

  c Dendrites: fine, thread-like extensions of the neuron that convert external signals to nerve impulses within the neuron

  d Myelin: the fatty layers of insulation surrounding the axon of a neuron

2 Schwann cells wrap their elongated, flat cells around and around the axon. The cytoplasm of the Schwann cells gets squashed into a small corner of the cell with all of the organelles, so the part of the cell wrapped around the axon is only a layer of membranes.

3 Students' answers will vary but should focus on answering both parts of the question. For the first part, students should remember that all of the sensory nerves for the whole leg are bundled into one nerve fibre, which, if compressed, would not be able to transmit signals to the CNS about the environment in and around the leg. If the brain is not receiving any information about the condition or position of the leg, it appears to be numb. For the second parts, students should recall that nociceptors (pain receptors) respond only to intense stimuli. Given that no signals at all have been received for some time, the first signals to arrive once the nerve fibre is released appear to be intense relative to previous signal levels, triggering the nociceptors to report pain.

■ 10.2 COMPARING THREE TYPES OF NEURONS

1

SIMILARITIES	DIFFERENCES
Examples may include figurative structural similarities, such as having a cell body, axon and dendrites and they may both be wrapped in a myelin sheath.	Students should think of many more differences than similarities, such as cells are living and the model is made of pipecleaners, the model is fluffy, it doesn't have the ability to transmit a signal and it doesn't have neurotransmitters and ions inside it.

9780170411660

2

Sensory neurons
- Soma is to the side of axon
- Dendrites are receptors
- Axon terminals link to CNS
- Transmit impulses towards CNS
- Detect stimuli

Elongated axon
Myelin sheath with nodes of Ranvier
Stretch the length of outer limbs

Motor neurons
- Soma is at the beginning of axon
- Dendrites link to the CNS
- Axon terminals link to muscles and glands
- Transmit impulses away from the CNS
- Triggers responses to stimuli

■ 10.3 PASSAGE OF A NERVE IMPULSE

The modified model should include a longer axon to show multiple jumps of the signal. The myelin sheath would need to run along the axon membrane, with small gaps for the nodes of Ranvier. The sodium ions should be grouped at the nodes, while potassium is spread evenly throughout the axon. When the action potential is triggered at one node, the sodium inside diffuses along the axon until it triggers more sodium to enter at the next node and diffuses along to trigger even more sodium at the next, etc.

■ 10.4 TRANSMISSION AND TRANSDUCTION

1

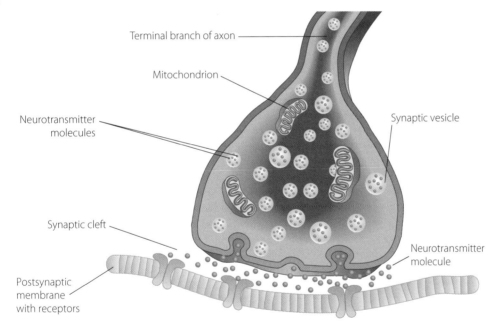

Terminal branch of axon

Mitochondrion

Neurotransmitter molecules

Synaptic vesicle

Synaptic cleft

Neurotransmitter molecule

Postsynaptic membrane with receptors

2

Signal transmission
- Alternating action and resting potentials
- Within a single neuron
- Increased by nodes of Ranvier
- Driven by ion movement

Sending a nerve impulse
Initially triggered by ions

Signal transduction
- Between two neurons, across the synapse
- Driven by neurotransmitters
- Opens ion channels in the receiving neuron

3 Students' answers will vary but should consider that the entire nervous system is interlinked and any signal could theoretically reach any other neuron in the body. If all neurons were directly connected, the action potential would propagate throughout the whole body without control. The system only works because of the pathways formed and that requires the signal to pass between specially selected neurons. Neurons that don't share a synapse are not meant to share signals.

CHAPTER 10 EVALUATION

1 A

2 D

3 • First, the sensory neuron detects the stimulus.

• Then the interneurons process the signal.

• Then the motor neuron triggers the response in an effector.

4 Dendrites are the first part of the neuron to receive a signal and pass it on down the axon. The axon terminals are the last part of the neuron to receive the signal and they pass the signal on to another neuron or an effector.

5 The impulse begins with external stimulation, which opens ion channels in the membrane. This allows sodium ions into the neuron, causing an action potential. As the sodium diffuses down the axon from an area of high concentration to the area of low concentration, it meets a node of Ranvier, where it triggers the depolarisation of the membrane and even more sodium flows into the neuron. This pushes the impulse down the axon faster, with each node it meets. As the wave of sodium ions moves off down the axon, the sodium-potassium pumps in the membrane pump the remaining sodium out of the axon, restoring resting potential.

6 Students' answers should focus on the structural differences (soma, dendrites and axon terminals), structural similarities (axon, myelin, nodes of Ranvier and length) and functional differences (direction of impulse, dendritic and axon linkages and role in responses to stimuli).

CHAPTER 11 REVISION

■ 11.1 HORMONES AS CHEMICAL MESSENGERS

1

GLAND	HORMONE	TARGET EFFECT
Testes	Testosterone	Regulation of aggression and competitiveness
Pancreatic alpha cells	Glucagon	Release of glucose from the liver
Pituitary	Antidiuretic hormone	Reabsorption of water in the kidney
Adrenal	Adrenaline	Increased blood pressure by vessel constriction
Pancreatic Beta cells	Insulin	Uptake of glucose from the blood
Thyroid	Thyroxine	Increased metabolic rate in almost all tissues

2 Testosterone is the only hormone from Question 1 that is a steroid. Students may justify this on the basis of the target effect that it is involved in aggression and competitiveness, but should also mention increased muscle mass or bone density as benefits in the sporting arena that steroids such as testosterone may have.

■ 11.2 UPREGULATION AND DOWNREGULATION

1 a Upregulation: when a cell is prompted to produce more of a particular cellular component, such as enzymes or receptors

 b Downregulation: when a cell is prompted to produce less of a particular cellular component, such as enzymes or receptors

2 Students' answers will vary but may include short-term periods of need (such as storage of glucose after a meal) or long-term periods of development (such as puberty).

3 Homeostasis maintains body conditions within narrow limits through negative feedback. Negative feedback is when the response counteracts or suppresses the stimulus. In this sense, downregulation (the suppression of gene products) would be a useful response mechanism in negative feedback

9780170411660

1

Fat-soluble hormone binding	Water-soluble hormone binding
The hormone passes into the cell before binding with a receptor inside – this indicates that the hormone can move freely across cell membranes and thus must be fat-soluble. The hormone–receptor complex also has a direct effect on the cell, without the use of secondary messengers.	The hormone never enters the cell itself. Instead, it binds to a receptor on the cell surface – this indicates that it cannot pass through the cell membrane and must not be fat-soluble. It also activates a cascade of secondary messengers that carry out the cellular response on behalf of the hormone, which is a key feature of water-soluble hormone activity.

2 The lock-and-key model states that hormones and their receptors are structurally compatible and that the hormone does not bind to other receptors because their 'locks' are not specifically designed to fit the exact 'key' of the hormone. One limitation of this model is that the static nature of the 'lock' does not account for the fact that the same receptor can be inactive or active depending on the interactions of other molecules or conditions. To address this, the induced fit model suggests the 'resting' conformation of a receptor is not specific to the hormone and can be deactivated by certain molecules or conditions and that the act of the hormone binding pulls the receptor into the correct, specific shape for that hormone.

CHAPTER 11 EVALUATION

1 C

2 A

3 Fat-soluble hormones can pass through cell membranes to bind with receptors internally. Often, the receptor–hormone complex alters cellular activity itself. Water-soluble hormones cannot pass through cell membranes and need to bind to receptors on the cell surface, which activates secondary messengers within the cell to alter cellular activity on behalf of the hormone.

4 Upregulation in hormonal control pathways refers to the process of a hormone triggering the mass-production of its own receptor.

5 For a small amount of hormone to have a large effect inside a cell, the signal must be amplified. This occurs when the receptor–hormone complex (either in the cell membrane or in the cytoplasm) activates several of the first of the secondary messengers. The first messsenger in the signalling cascade activates several of the second messenger, which activates several of the third messenger. With enough messengers in the cascade, the signal that began with one hormone–receptor complex becomes amplified to hundreds of effector molecules carrying out the cellular response.

6 Students' answers will vary but could include constricting blood vessels to enable faster blood flow and the release of glucose from the liver. Cell targeting works by having the same hormone bind to different receptors in different cells. The receptor for adrenaline in liver cells activates a signalling cascade that results in the bulk release of glucose from stored glycogen, while the receptor for adrenaline in blood vessel cells activates a different cascade that results in constriction of the vessels and other cells in the body have no receptor s for adrenaline at all and therefore, do not respond to it.

CHAPTER 12 REVISION

1 The increase in metabolic rate generates more heat, assisting to maintain body temperature at the ideal range. The body is responding to the colder environmental temperature by increasing metabolic rate.

2 Increase activity levels; for example, by running (voluntary response/behaviours), shivering (involuntary muscle contractions) etc.

3 Just below 30°C (around 27°C) to just above 35°C (around 37°C).

4 Some chemicals, such as enzymes, may not be able to function outside the critical temperatures, so chemical reactions in the cells may not proceed at the necessary rate to maintain normal functions.

5 The process is hibernation. During hibernation, the metabolic rate falls to a level that just sustains life; the set point is lowered considerably.

6 Aestivation

1 Groups of penguins can huddle together, forming large 'circles' – the birds on the outside of the group protecting those in the inner part from the strong cold winds, losing less heat to the environment. Individuals take turns to move from the outside of the huddle to a more protected inner layer, to maintain stable body temperatures.

2 Emperor penguins have a very effective system of keeping their extremities warm –counter-current heat exchange. Blood travelling in the arteries to the foot warms the blood returning to the body in the adjacent veins. The outgoing blood to the extremity is cooled in the process but not enough to affect cell activities. As the temperature gradient between the extremity and the surroundings is reduced, heat loss is minimised.

3 Emperor penguins are well insulated by several layers of scale-like feathers, so they lose less heat to the environment than other animals that do not have this structural feature.

■ 12.3 STRATEGIES AND ADAPTATIONS FOR THERMOREGULATION

Students' answers will vary but can include the following descriptions and examples.

Structural features:

- Insulation: a layer or layers of a material that surround an organism or part of an organism reducing heat loss or heat gain. Examples include feathers and fur.
- Brown adipose tissue: a type of fatty tissue richly endowed with blood vessels and mitochondria. It insulates and helps generate metabolic heat. Examples include high levels in hibernating animals and newborn humans.
- Shape and size: a small surface-area-to-volume ratio reduces heat loss and vice versa. Examples include larger animals being found in cooler climates.

Behavioural responses:

- Moving locations: moving from one area to another. Examples include moving into the shade or into burrows to reduce heat gain in hot temperatures.
- Huddling: animals in a group move closer together. Examples include penguins huddling together to reduce exposure to the cold and sharing body warmth.
- Torpor: a state of decreased activity and metabolism that allows animals to survive unfavourable conditions and/or conserve energy. Hibernation and aestivation are states of torpor. Examples include Australian carnivorous marsupials burrowing and going into torpor during the hottest part of the day and animals hibernating in winter.

Physiological mechanisms:

- Vasomotor control: altering the amount of blood flow near the surface of the body. Examples include vasodilation and vasoconstriction.
- Evaporative heat loss: evaporation of water from the moist skin cools the blood as it flows through capillaries near the surface. Examples include sweating, panting and licking external areas.
- Counter-current heat exchange: arteries and veins located close to each other in the heat exchanger allows blood travelling in the arteries to the foot or fin warming the blood returning to the body in the adjacent veins. Examples include feet of penguins.
- Thermogenesis: the process of heat production in organisms. Examples include shivering.

Homeostatic mechanisms:

- Hormone negative feedback: the process of maintaining a condition of equilibrium or stability within the internal environment when dealing with external changes. Examples include thyroid and insulin hormones.

■ 12.4 BILBY THERMOREGULATION

1

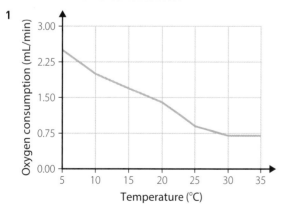

2 At particular external temperatures, the metabolic rate of an animal begins to rise by increasing aerobic cellular respiration. Oxygen is consumed in this process. This increases heat output.

3 The external temperature at which the metabolic rate begins to rise is the lower critical temperature. This is between 25°C and 30°C for the bilby.

4 Oxygen consumption did not vary at 30°C and 35°C because this is the optimum temperature range.

5 Huddling reduces overall heat loss by reducing the group's surface-area-to-volume ratio. Body warmth between bilbies may also be shared.

6 Vasoconstriction and shivering

7 Using 16 bilbies is more accurate. For example, one bilby may have atypical responses or may be sick. If more bilbies are used, any difference in results will be considered and if one result, is different from the other results, it could be discarded. Increasing the sample size improves reliability of results.

CHAPTER 12 EVALUATION

1 C

2 Students' answers will vary. Physiological adaptations include vasomotor control, evaporative heat loss, counter-current heat exchange and thermogenesis. Behavioural adaptations include moving locations, huddling and torpor.

3 a Hypothermia

b Students' answers will vary but can include wearing warm, protective clothing, keeping dry and sheltered, and drinking warm liquids.

4 a Thyroxine. The hypothalamus would release thyrotropin-releasing hormone, which activates the anterior pituitary gland to release thyroid-stimulating hormone (TSH). TSH stimulates the thyroid gland to secrete the hormone thyroxine into the blood.

b Thyroxine increases the metabolic rate, which increases heat production especially in the liver.

c When temperatures start to increase again, the hypothalamus responds by reducing the release of TSH by the anterior pituitary gland so less thyroxine is released from the thyroid gland. Thyroxine itself has an inhibitory effect on the anterior pituitary gland, which responds by secreting less TSH when thyroxine levels are high.

5 a Brown adipose tissue has more blood vessels and mitochondria than white adipose tissue.

b In human infants, brown adipose tissue makes up about 5% of total body weight. Newborns do not shiver in the cold and are poorly insulated. Brown adipose tissue is a valuable adaptation that generates heat when they are cold. The brown adipose tissue gradually reduces with age.

CHAPTER 13 REVISION

13.1 WATER BALANCE IN ANIMALS

Students' answers will vary but may include the following.

Structural:

- Bushy tail: the bushy tail can be wrapped around the face. This strategy reduces water loss by saturating the air between the hairs at its body surface and the air in the burrow with water vapour.
- Loop of Henle: water is conserved when it is removed in the descending portion of the loop of Henle. The longer the loop of Henle, the more concentrated the urine and the more water saved. The spinifex hopping mouse has a very long loop of Henle to maximise water conservation.

Behavioural:

- Nocturnal activity: enables the mouse to avoid daytime extremes in temperature hence reducing water loss.
- Burrowing: areas below the surface are significantly cooler, have higher humidity, experience no sunlight, and also experience lower levels of infrared radiation. All of these conditions help to reduce water loss.

Physiological:

- Concentrated urine: the ability to concentrate urine reduces the amount of water that is lost from the mouse when metabolic waste is released. The spinifex hopping mouse produces the most concentrated urine known for any mammals and also very dry faeces.

Homeostatic:

- Antidiuretic hormone (ADH): this hormone is responsible for increased permeability of the distal tubules of the kidney, increasing water reabsorption and reducing urine volume. High levels of ADH reduce water loss in spinifex hopping mice.

13.2 WATER BALANCE IN PLANTS

1 Plant A is a hydrophyte.

Plant B is a xerophyte.

Plant C is a mesophyte.

2 Plant A has stomata on the upper surface. Plant B has sunken stomata on the lower surface surrounded by hairs. Plant C has stomata on the lower surface.

3 Plant A lives in a wet environment. These plants do not need adaptations for water retention. The leaf has an abundance of air-filled intercellular spaces because water contains much less oxygen than air. Stomata are seen on the upper surface of the leaf for gas exchange into the air.

Plant B lives in a dry environment. There is a thick cuticle on the upper surface, reducing water loss. The stomata on the lower surface are in sunken pits surrounded by hairs, which creates a humid microclimate, reducing transpiration.

Plant C lives in a moderate environment. The leaf has a thinner cuticle than a xerophyte with stomata located on the undersides of the leaves. The shape indicates the leaf is flat. These features reduce water loss but not to the extent of a plant in a dry environment.

4 Students' drawings will vary. Answer may include drawing and explanation of features such as thick waxy cuticle on the leaf surface, reduced numbers of stomata on the top of the leaf and increased numbers on the bottom of the leaf, sunken stomata, cylindrical or rolled leaves, reduced leaf numbers or no leaves, hairs on leaves, large vacuoles to store water, long vertical roots or superficial horizontal roots.

13.3 COMPARING OSMOREGULATION IN DIFFERENT HABITATS

1 a Mammal B

b Less water gained through drinking because less water available.

Less water lost through evaporation because adaptations would be used to reduce water loss.

Less water lost in urine. Water retention adaptations would be present.

Higher ADH produced so less water lost in urine.

2 a Pituitary gland

b Higher levels of ADH in mammal B would increase the permeability of cells lining the collecting ducts, facilitating the osmotic movement of water into the surrounding blood. As a result, less urine is produced and the urine is more concentrated compared to mammal A.

3 Mammal B would have a longer loop of Henle than mammal A.

CHAPTER 13 EVALUATION

1 A

2 Osmoconformers have body fluids isotonic to the surrounding water.

3 Large vacuoles occupying most of the cell volume.

4 An animal could be nocturnal and spend time in a burrow.

5 a The thick cuticle restricts water loss and at the same time allows light rays to pass through into photosynthetic parts of the plant. Sunken stomata create a humid microclimate, reducing transpiration and hence water loss.

b Reduced numbers of stomata on the top of the leaf and increased numbers on the bottom of the leaf, cylindrical or rolled leaves, reduced leaf numbers or no leaves, hairs on leaves, large vacuoles to store water, long vertical roots or superficial horizontal roots.

6 Leaves have an abundance of air-filled intercellular spaces as water contains much less oxygen than air. Stomata are seen on the upper surface of the leaf for gas exchange into the air.

CHAPTER 14 REVISION

14.1 INFECTIOUS AND NON-INFECTIOUS DISEASES

1 Infectious diseases are caused by any agent (pathogen) that can be transmitted from one organism to another. Non-infectious diseases are not transmitted between individuals and include nutritional and genetic diseases.

2

DESCRIPTION OF DISEASE	INFECTIOUS	NON-INFECTIOUS
Uncontrolled growth and division of lung cells		√
This disease can be prevented by boiling water before drinking it	√	
Reproduction of particles inside skin cells before escaping and entering other skin cells	√	
Proteins that convert the normal form of a protein to a harmful form, which can then convert more normal to abnormal forms	√	
Rise in blood glucose after consumption of sucrose and glucose		√

14.2 VIRUSES AND BACTERIA

1 1 Virus particle binds to the wall of a suitable host cell and viral DNA enters the cell's cytoplasm.

2 Viral DNA directs host cell machinery to produce viral proteins and copies of viral DNA.

3 Viral proteins are assembled into coats; DNA is packaged inside.

4 Tail fibres and other components are added to coats.

5 Host cell undergoes lysis and dies. Infectious virus particles are released.

2 1 Pilus

2 Cell wall

3 Cytoplasm

4 Plasma membrane

5 Bacterial flagellum

6 Ribosome

7 DNA

8 Capsule

14.3 TRANSMISSION OF DISEASE

1

DISEASE	PATHOGEN	METHOD OF TRANSMISSION
Common cold, influenza	Virus	Contact with body fluids via coughs and sneezes, and wiping nose and touching others
Tinea (athlete's foot)	Fungus	Direct contact
Typhoid	Bacteria	Contaminated water
Bubonic plague	Bacteria	Insect vector – flea
Gonorrhea	Bacteria	Contact with body fluids
Rotavirus	Virus	Contaminated food
Cold sores	Virus (*Herpes simplex*)	Direct contact
Salmonella food poisoning	Bacteria	Contaminated food

2 a Malaria is transmitted by the *Anopheles* mosquito vector. Zygotes of the malarial pathogen, *Plasmodium*, develop in the gut of the female mosquito then migrate to her salivary glands. When the mosquito bites a person, the pathogen is injected into their bloodstream.

b Reduce availability of breeding areas for mosquitoes by draining swamps and covering the water surface with kerosene which kills mosquito larvae. Kill adult mosquitoes by spraying air and walls of homes with insecticide. Sleep under mosquito nets.

c Malaria can be transmitted by blood transfusion because donated blood may contain merozoites that result from asexual reproduction in the host. For this reason, people who have travelled to countries where malaria occurs may be deferred from giving blood for a short period. Malaria can also be transmitted from a mother to her foetus.

d *Plasmodium* is a protist.

3 a Schistosomiasis is a waterborne disease caused by the South-East Asian blood fluke, *Schistosoma japonicum*. Inside the human host, the blood flukes reproduce sexually and their fertilised eggs leave the human body in faeces. On contact with fresh water, they hatch into ciliated swimming larvae that burrow into snails where they multiply asexually. The fork-tailed larvae leave the snail and bore into human skin.

b The aquatic snail is the intermediate host.

c The human is the definitive host.

d Control of schistosomiasis could be based on improved sanitation, health education and snail control that includes draining swamps because the flukes depend on water to complete their life cycles.

e Having the snail involved in the life cycle as the intermediate host means that asexual reproduction can build up numbers of larvae and the larvae in the snail form a reservoir of larvae which can hatch out intermittently in search of a host. Both of these adaptations increase their chance of transmission.

■ **14.4 ADAPTATIONS TO ENSURE TRANSMISSION OF PATHOGENS TO NEW HOSTS**

1 Irritation of the mucous membranes lining the nose and throat make the host both:

- secrete large amounts of mucus (body fluid), which they will wipe away with their hands, ready to infect other people they touch
- cough and sneeze mucus out of their bodies, propelling their body fluids out towards other potential hosts.

2 • Asymptomatic shedding of *Herpes* virus on the skin between the occurrence of visible sores means the person will not know they are infective, so will not take precautions against infecting others.

- When visible sores are very itchy, the host will scratch the sore, releasing virus onto their skin and under their fingernails from where it can easily spread to other hosts.

3 Symptoms of messy, watery diarrhoea containing 10 000 million (that is 10^{10}) virus particles per millilitre of faeces means that when a person visits the toilet or changes the nappy of an infected infant, their hands can easily become contaminated with pathogens. As the infective dose is only 100 to 10 000 virus particles it is easy for enough pathogens to be transmitted via food unless very strict handwashing procedures are followed.

4 The itchiness caused by the batch of eggs on the skin surrounding the anus will induce the host to scratch the region. They will scrape eggs under their fingernails and onto their hands, leading the host to re-infect themselves or others by transferring eggs to their mouths or onto food.

5 The migration to and reproduction of the bacteria in the stomach, where it forms a plug are two important features.

- The plug means that when the flea bites, it vomits blood tainted with the bacteria back into the bite wound, passing on the plague.
- It also causes hunger, making the flea first bite its host rat voraciously, then look for other hosts to feed from when the rat dies of the disease.

6 Being able to reproduce in and on food is essential for bacteria like *Salmonella* because even though only a small dose may contaminate food, the infective dose of $>10^5$ (100 000) can be reached in a short time by reproduction. This makes it important to keep food out of the temperature danger zone of between 5°C and 60°C, where most bacteria can grow and reproduce.

■ **14.5 VIRULENCE FACTORS**

1 a Pathogenicity is the disease-causing capacity of a pathogen.

b Virulence is the intensity of the effect of the pathogen on its host. It is a measure of a pathogen's ability to cause disease within its host.

2 Virulence factors are characteristics that promote the establishment and maintenance of disease. They help pathogens to invade the host, cause disease, and evade host defences.

9780170411660

3

CHARACTERISTIC	EXOTOXIN	ENDOTOXIN
Relative toxicity	More toxic than endotoxins	Less toxic than exotoxins
Chemical make-up	Protein toxins and enzymes	Lipopolysaccharide – a complex of lipid and polysaccharide
Source	Actively secreted by pathogenic bacteria into their surroundings	Attached to the outer membrane of certain bacteria and released upon lysis of the bacterial cell
Nature of their action	Act locally or in tissues distant from the site of bacterial growth. Major categories include cytotoxins, neurotoxins, and enterotoxins	Cause fever, changes in blood pressure, inflammation, lethal shock, and many other toxic events
End result	Among the most toxic of all substances, may lead to death	Can culminate in sepsis and may lead to death

4 Virulence factors that facilitate *adherence* (binding to host cell surfaces) are known as *adhesins*. Adherence factors ensure that the pathogen attaches to the *tissue* type in which it can survive. Many pathogenic bacteria recognise and attach to epithelial surfaces by using *pili*. These are fine filaments of *protein*, up to several *micrometres* in length and resembling fine hairs. Adhesins also include a wide variety of other *surface* proteins, as well as bacterial cell *walls* and bacterial *capsules*.

Invasion factors are virulence factors that facilitate bacterial *invasion* of a host. They enable *entry* into the cells and tissues of the host in order to ensure its *colonisation*. Invasion factors are often *enzymes* secreted by bacteria. One such *enzyme* degrades a structural component of *blood* clots, facilitating bacterial transport across *epithelial* layers and penetration into the *surrounding* tissues. A successful invasion by intracellular pathogens means penetrating host cell *membranes*. Surface proteins found on some *bacteria* allow them to invade mammalian cells via *transmembrane* proteins.

Lifestyle changes, for example *spore* formation, can result in increased *pathogenicity*. The *bacterium* that causes tetanus can last for years in soil as an *inert* spore that will resume growth when conditions become more favourable inside a new host.

The *capsule* is a large, well-organised layer made of thick *polysaccharide* gel that forms part of the outer *structure* of many bacterial cells. Encapsulated strains of *bacteria* have been shown to be more *virulent* than non-encapsulated strains, apparently because capsules inhibit *phagocytosis* by host phagocytes.

14.6 SECOND-HAND DATA ANALYSIS: THE EFFECTIVENESS OF ANTIMICROBIALS

1 • The bench was wiped down with bleach and participants washed their hands thoroughly. This was to ensure that participants did not ingest bacteria and that there would be no cross contamination with bacteria in the future.

   • The plates were sealed with sticky tape. This prevents the escape of potentially pathogenic bacteria into the environment, from where they could infect a person.

2 The disc dipped in water was the control. It showed how much of the zone of inhibition (the clear area) was due to a liquid washing the bacteria off the agar, rather than being killed by an antibacterial.

3 Using three agar plates provides three trials that can be averaged. This reduces the effect of random errors, improving the reliability of the data.

4

	LYSOZYME	ANTISEPTIC	DISINFECTANT	WATER
Mean	13	14	15	7

5 Independent variable: type of antibacterial solution

   Dependent variable: diameter of the zone of inhibition, the clear area around each disc, showing the degree of sensitivity of the bacteria to each substance

6 Order of effectiveness, from most to least: disinfectant, antiseptic, lysozyme

7 The results for the three trials show that the results are precise. This suggests good control of extraneous variables.

1 C

2 **a** African sleeping sickness is transmitted by the tsetse fly vector. This aids transmission because the fly can carry the disease considerable distances and between suitable hosts. Asexual reproduction in the tsetse fly also increases the number of infective forms of the pathogen.

  b This disease could be reduced by removing cows from around where people live, killing the tsetse flies and destroying the habitat in which they live and reproduce.

  c Having both sexual and asexual reproduction means that large numbers of infective forms of the pathogen can build up, making transmission more likely.

3 Any two differences:

Exotoxins are:

- more toxic than endotoxins
- protein toxins not lipopolysaccharide
- actively secreted by pathogenic bacteria not part of the structure of certain bacteria and only released upon lysis of the bacterial cell
- cytotoxins, neurotoxins, and enterotoxins not stimulants of the immune system causing fever, changes in blood pressure, inflammation, lethal shock, and many other toxic events

Any two similarities:

Exotoxins and endotoxins are both:

- toxic to the host
- products of bacteria
- causes of serious illness or death
- virulence factors

CHAPTER 15 REVISION

15.1 DETECTION OF INVADERS

1 'Self' refers to substances and cells that belong in any particular organism; 'non-self' refers to what is foreign. A substance identified as 'non-self' is likely to stimulate an immune response.

2 An antigen is any substance that triggers an immune response.

3 Both plants and animals are alerted to the invasion of bacteria and viruses by *physical* and *chemical* changes that occur in their cells or tissues. Antigens are generally protein or polysaccharide molecules *foreign* to the host. Their presence, either on the outer *surface* of the invaders or in the *toxins* and enzymes they secrete, stimulates host *immune* responses that usually lead to the *destruction* and removal of the pathogen.

Pattern *recognition* receptors are proteins used by nearly all organisms to identify molecules associated with pathogens. These *receptors* are commonly found on the *surface* and in the *cytoplasm* of host body cells. They recognise specific *substances* and *molecular* patterns that are characteristic of a wide *variety* of pathogens, but not found on *host* cells. These molecules are called pathogen-associated molecular patterns and include lipopolysaccharides, glycoproteins and particular protein sequences on the *surface* of invaders. Even a small part of a molecule, called an *epitope*, may be antigenic.

A particular receptor can recognise a variety of different pathogens if all of them display the same *molecular* pattern. For example, the material that makes up bacterial flagella, called *flagellin*, is found in a wide variety of bacteria. This enables a *flagellin* receptor to recognise many different types of *bacteria* as invaders. This system of *recognition* has the advantage of activating a *rapid* response to invaders but it lacks a high degree of *specificity*.

15.2 INNATE AND ADAPTIVE IMMUNE RESPONSES

Activity 1

1 1 Bacteriostatic substances in ear secretions

  2 Mucus and cilia trap and remove debris and microscopic pathogens

  3 Secretions in sebaceous glands

  4 pH and commensal organisms in vagina, antibacterial proteins in semen

  5 Continual washing (tears and protective enzymes)

6 Acid in sweat

7 Skin

8 Stomach acid and commensal organisms

9 Passing urine flushes urethra

10 Blood clotting and wound repair

2 1 Hair on leaf surface may help deter pathogens.

2 Waxy cuticle prevents entry of many pathogens.

3 Hairs may help prevent entry of pathogen through stomata.

Activity 2

1 False

2 True

3 True

4 False

5 False

6 True

7 False

8 False

9 True

10 False

▪ 15.3 RESPONSE OF A BODY UNDER ATTACK

1 a 3

b 11

c 4

d 5

e 6

f 7

g 9

h 10

i 1

j 8

k 2

2 Innate responses: i, k, a, c, b.

▪ 15.4 ACTIONS OF LYMPHOCYTES

Activity 1

	B LYMPHOCYTES	HELPER T (T_H) LYMPHOCYTES	CYTOTOXIC T (T_c) LYMPHOCYTES
Development of self-tolerance	Occurs in bone marrow	Occurs in thymus	Occurs in thymus
Undergo clonal selection (Yes/No)	Yes	Yes	Yes
Effector functions	Plasma cells produce antibodies	Production of cytokines to aid B cell, T_c cell and macrophage functions	Destruction of virally infected and cancerous cells
Formation of memory cells (Yes/No)	Yes	Yes	Yes

2 **a** Complement activation: antibodies that are bound to antigens are potent activators of the complement cascade.

   b Opsonisation: bound antibodies are able to attract phagocytes, effectively 'tagging' pathogens for phagocytosis and destruction.

   c Neutralisation: some antigens can act as toxins and cause cellular damage. In these cases, antibodies neutralise toxins by preventing them from binding to their target.

   d Agglutination: the binding of antibodies can immobile pathogens because they become stuck together in an antibody–pathogen net. This means the pathogens are not able to spread. Being clumped together in one spot also makes them more susceptible to destruction by phagocytosis.

Activity 2

1 The immune system rejects transplanted organs because it recognises the cells and tissues as foreign, due to their 'non-self' surface antigens.

2 The patient will become more susceptible to viral diseases and tumours. Also, rapidly dividing cells such as those in the intestine may be affected, causing nausea and intestinal disturbances.

3 Cytotoxic T (T_C) cells, activated by encountering the transplanted organ, will divide many times to form a group of clones specific for the transplanted cells. Some become effector cells that are effective killers, releasing powerful cytotoxins that induce apoptosis in any transplanted cells and tissues. If DNA synthesis is prevented, this sequence of events cannot occur, effectively limiting transplant rejection.

4 No; he should not be concerned. Surface molecules on body cells that identify them as 'self' are genetically determined, meaning that identical twins will recognise each other's organs as 'self' and will not reject them. He therefore does not need immunosuppressant drugs.

■ 15.5 ANTIBODIES IN ACTION

Activity 1

	ACTIVE	PASSIVE
Naturally occurring	Exposure to a pathogen	1 Transfer of antibodies from mother to foetus via the placenta 2 Transfer of antibodies from mother to baby via breast milk
Artificial	Vaccination	1 Anti-venom – antibodies against spider or snake venom 2 Antibodies against particular pathogens, e.g. rabies

Activity 2

1 Antibodies

2 Plasma cells

3 Once in the horse, the venom would bind to complementary B cell receptors (antibodies) on the surface of some B cells. These activated B cells would divide rapidly to form effector and memory B cells. The effector B cells, called plasma cells, secrete huge quantities of antibodies, specific for that venom. The memory cells migrate to the lymph nodes. After another dose of venom these memory cells divide rapidly to form large numbers of plasma cells that secrete even larger quantities of antibody. Memory cells are also produced and remain ready to respond to the next injection of venom, which will produce larger and larger quantities of antibodies each time.

4 Injecting small amounts of venom into the horse will not negatively affect the horse. However, over a long period of time, both the amount of antibody and the number of memory cells will gradually increase so that very large quantities of antibodies can be harvested from the horse blood after 10–12 months.

5 Anti-venom

6 The antibodies rapidly bind to the snake venom. This causes agglutination (clumping of the venom), which prevents it from acting as a neurotoxin.

1 A

2 a Any two of:

- Tannins in bark and some leaves are toxic to insects and their bitter taste deters herbivores.
- Caffeine is an alkaloid found in plants such as coffee, tea and cocoa that is toxic to both insects and fungi.
- Pyrethrins are compounds produced by chrysanthemum flowers that act as insect neurotoxins.
- Asparagus plants and marigolds secrete toxins into the soil that kill nematodes.
- Plant oils, for example tea tree oil, eucalyptus oil and lavender oil, function as insect toxins and protect against fungal or bacterial attack.

b Defensins are small, stable a small antimicrobial peptides secreted by virtually all plants and often present in flowers, leaves, bark, fruit and seeds.

Any two of:

- They are able to provide a defence against insect-transmitted viruses by stopping feeding by insects.
- They are able to inhibit the growth and development of fungi, as well as bacteria, viruses and insects.
- They are antimicrobial in action because they inhibit the action of a pathogen's enzymes and ribosomes.

3 Fewer ciliated cells in the airway epithelium and decreased ciliary beat frequency caused by cigarette smoke would prevent effective removal of foreign substances and bacteria trapped in mucus in the lungs. Trapped bacteria may grow and reproduce in the mucus, setting up serious infections in the lungs. Foreign substances could include irritants and carcinogens, which could cause lung problems like asthma, infections and cancer.

4 a • Argument that refutes 'Warfarin is a blood thinner':

Warfarin does not thin the blood. It works by reducing the amount of clotting protein in the blood because it reduces the liver's ability to use vitamin K to make these blood-clotting proteins. With less clotting protein, there will be less clotting, making a person prone to bleeding.

- Argument that supports 'Warfarin is a blood thinner':

Warfarin works by reducing the amount of clotting protein in the blood because it reduces the liver's ability to use vitamin K to make these blood-clotting proteins. With less clotting protein, there will be less clotting, making a person prone to bleeding. The person appears to have thinner blood because they bleed and bruise easily.

b A diet very low in green vegetables will also reduce the amount of clotting protein in the blood as green vegetables provide vitamin K, which the liver uses to make clotting proteins. These people will bleed and bruise more easily.

5 a Platelets are cell fragments that assist blood clotting and wound repair, preventing the entry of microorganisms into the body.

b Anti-platelet antibodies would bind to platelets, rendering them ineffective in clotting the blood. This would lead to the symptoms of bruising, rashes and internal bleeding.

c Platelets from blood donations would not be effective at preventing symptoms in patients with ITP because their anti-platelet antibodies would attack and destroy the extra platelets in the same way.

16.1 DESCRIBING DISEASE

1 Students' answers will vary. For example: Tasmanian devils are suffering from a new disease that is decimating the population. Devil facial tumour disease (DFTD) is an infectious disease that emerged in the mid-1990s. Scientists were initially baffled by the disease, taking some time to work out that DFTD is a very unusual and very rare contagious cancer. Cancer is not usually something you catch; it is something that dies with its host.

In this cancer, the tumour cells themselves are the infectious agent and they are spread from devil to devil by biting, especially during mating. Once in the new host, the cancer cells grow into new tumours that block the eyes, mouths and ears, causing the animals to starve to death. This devastating disease has brought about devil population declines in excess of 90% in many areas of Tasmania.

2

CONCEPT	DEFINITION
Epidemic	A very serious increase in the occurrence of a particular disease above the baseline level for that population
Endemic	A disease that is prevalent at a constant rate within a population
Morbidity	The impact of a disease within a population, measured by the number of people affected by that disease
Outbreak	An increase in the occurrence of a particular disease above the baseline level for that population
Mortality	The impact of a disease within a population, measured by the number of deaths caused by that disease
Virulence	The ability of a pathogen to cause severe disease within its host
Infectivity	The ability of a pathogen to spread from one host and infect another host

■ 16.2 ANALYSING THE CONSEQUENCES OF INFECTION

1 Virus

2 Direct contact; body fluids; foodborne transmission; waterborne transmission; disease-specific vector

3 Any two of:
- whether the susceptible individual was immunised against the disease
- whether the individual had been previously infected by the disease; in both cases they would have enough antibodies to prevent reinfection
- whether they had been exposed to a high enough dose of the pathogen to establish an infection

4 • Resolution of infection: infected person had enough natural resistance to eradicate the pathogen before showing symptoms.
- Symptomatic infection: pathogen infects person who shows symptoms of the disease.
- Three possibilities for symptomatic person:
  - Their immune system effectively fights the disease and they resolve the infection.
  - Their immune system is unable to effectively fight the disease and they die from the infection.
  - The person is able to control the infection to some degree, but is unable to clear from the body, making them a long-term asymptomatic carrier of the disease.

5 Hepatitis, Epstein–Barr virus (causes glandular fever), HIV, poliomyelitis

6 Because they do not know they have the disease, asymptomatic carriers will not take any precautions against the spread of disease; nor will they seek treatment, yet they are still able to pass the disease on to others.

■ 16.3 CASE STUDY: SPANISH FLU

1 a Virulent: a pathogen that causes severe and intense disease in its host

b Infectivity: the ability of a pathogen to spread from one host and infect another host

c Susceptibility: the level of resistance of an organism to a pathogen

2 Exhaustion and poor nourishment reduce the effectiveness of the immune system in fighting disease. Living in close quarters and poor hygiene, which includes hand washing, would increase the likelihood of a pathogen, that is transmitted in body fluids, spreading through a population.

3 a Host: transportation of soldiers around the world, populations who were exhausted poorly nourished and living in close quarters as a result of the war had increased susceptibility to infection.

b Pathogen: this particular strain of influenza had high infectivity, was more deadly than regular seasonal influenza and caused many people's immune systems to overreact to infection.

c Environmental: war disrupted normal healthcare programs, leaving countries unprepared to respond and poor hygiene resulted from disruptions to housing and clean water.

4 A cytokine storm in the lungs causes fluids and immune cells such as macrophages to accumulate, blocking airways and causing suffocation.

5 SARS, swine flu, MERS and HIV-AIDS

9780170411660

16.4 DEFINING DISEASE

Quarantine: the enforced isolation of individuals at risk of carrying disease to prevent the spread of that disease into healthy populations

Intermediate host: an organism in which a pathogen or parasite undergoes development and spends a small proportion of its life cycle

Isolation: the enforced separation of individuals with a disease to prevent the spread of that disease into healthy populations

Definitive host: a host in which the adult phase of a parasite produces gametes

Carrier: an infected person able to control a disease to some degree, but still capable of transmitting infection to others

Vector: a living organism that transmits pathogens from one host to another

Epidemiology: the study of the causes and effects of diseases at a population level

16.5 OUTBREAK INVESTIGATION

Cloze activity

1 The first step is to confirm that the reported cases meet the definition of an *outbreak*. This involves comparing the number of diagnosed cases with *background* levels of the disease.

2 Next, investigators formulate a *case* definition. Case definitions include the type of illness, the place and the time.

3 Investigators then find people affected by the outbreak by using *contact* tracing; recent *contacts* of the *infected* person are traced and screened for the infection.

4 The type of *contacts* sought will vary with the *mechanism* of transmission. In a case of food poisoning, investigators focus on people *exposed* to the same food sources, not direct *contact* with an infected individual.

5 Investigators then gather *information* in interviews in which they ask about usual activities, sick contacts, recent meals and travel.

6 Once a *hypothesis* about how the outbreak is spreading has been generated, the investigators search for *evidence* to support or refute that hypothesis.

Case study

1 Case definition – type of illness: gastroenteritis caused by the norovirus. Place: a particular restaurant in an Australian coastal town. Time: one particular night in 2007.

2 Contact tracing: when recent contacts of the infected person are contacted and screened for the infection.

3 In this scenario, contact tracing meant that all of those who had dined at the restaurant on that night were contacted and asked about symptoms.

4 All of those who contracted the infection ate oysters for their entrée. No person who didn't fall ill ate the oysters. No evidence of the virus was found in the kitchen or in other batches at the oyster farm, so the investigators concluded the outbreak was likely caused by a single infected batch of oysters.

CHAPTER 16 EVALUATION

1 A

2 The movement of individuals and populations can facilitate the spread of disease. This is because individuals carrying disease are able to infect other individuals in the areas they are travelling to, allowing diseases to spread faster and over larger geographical areas than they otherwise could.

3 Epidemiologist

4 a Vector

   b Models suggest that increases in temperature and changes in rainfall are likely to result in the spread of *Anopheles* mosquitoes into previously uninhabitable regions, which would spread malaria more widely.

5 Herd immunity refers to the phenomenon that once a particular proportion of a population is immune to a disease; susceptible individuals are better protected from the disease. The disease cannot spread through the community because there are too few susceptible individuals.

6 a School and workplace closures: as centres of social activities, these places play an important role when disease transmission is by social contact.

b Reduction of mass gatherings: events such as concerts, religious observances, sports matches and festivals are characterised by a concentration of people in close contact with each other which potentially increases the risk of the spread of infectious diseases.

c Temperature screening: works by detecting infrared radiation. Different colours indicate body temperature. The idea is to detect and prevent a person with a fever-causing illness from entering a country, because of the risk that they will introduce a serious disease.

d Travel restrictions: the idea is to ban from entry all people from countries experiencing a disease epidemic, in the hope of preventing the disease spreading.

e Contact tracing: help to alert people to their infection before they experience symptoms. This ensures they take steps to avoid spreading the disease.

f Quarantine: stops exposed individuals from entering a healthy population until the incubation period of that disease has passed.

g Personal hygiene measures: regular handwashing, avoiding shaking hands, kissing and touching surfaces and objects that are used and shared by others prevents individuals from contracting infections that are spread by faecal–oral or direct contact routes. Keeping a metre away from people who are sick helps to prevent airborne transmission.

BIOLOGY UNITS 1 & 2

■ MULTIPLE-CHOICE QUESTIONS

1 A

Transport proteins act as passageways that allow specific substances to move across the membrane. They include carrier proteins that assist other molecules to cross the membrane in active transport and facilitated diffusion, pumps that require energy to carry out active transport and channel proteins that form narrow passageways to allow ions to passively cross the membrane, from high to low concentrations.

2 C

Membrane-bound organelles are an advantage because they partition the cell into specialised compartments. An organelle does not need to be membrane-bound to facilitate the synthesis or the breakdown of complex molecules.

3 B

At high temperatures, enzymes lose their functional shape and the substrate can no longer bind with the active site. If the shape has changed enough to break the bonds between the connecting units of amino acids, the change is permanent because the enzymes cannot return to their original shape when conditions revert to normal.

4 D

Plants carry out photosynthesis, in which light energy is converted into chemical energy. Both plants and animals convert chemical energy into heat energy in the process of aerobic respiration.

5 A

The constant movement of blood through the capillaries around each alveolus ensures the concentration of carbon dioxide remains higher in the blood than in the alveolus. This concentration gradient leads to the diffusion of carbon dioxide into the alveolus.

6 C

Water, urea, glucose, amino acids, and salt move out of the glomerulus into the Bowman's capsule by filtration. Glucose, amino acids and sodium ions are selectively reabsorbed from this filtrate in the proximal tubule, into the blood. Water then flows out of the tubule and into the blood by osmosis. Wastes such as antibiotics are secreted from the capillary network into the collecting duct.

7 B

Steroid hormones are hydrophobic so they are lipid soluble. They can pass through the plasma membrane, so their target cells are inside the cell.

8 A

The air in the depression around sunken stomata is protected from the wind, helping to keep high levels of humid air around the stomata. This reduces the rate of water loss from the leaf cells, enabling the stomata to remain open on hot, windy days.

9 B

After exposure to the influenza virus, lymphocytes were stimulated to produce and store memory B cells specific to the particular antigen of the influenza strain of that year.

10 D

Immunisation does not kill the flu particles and does not reduce the virulence of the virus. It will prevent the virus multiplying in an individual because they will have formed memory cells for that strain.

1 a

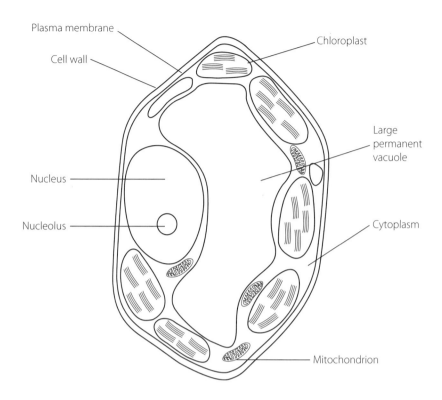

Plasma membrane

Cell wall

Chloroplast

Large permanent vacuole

Nucleus

Nucleolus

Cytoplasm

Mitochondrion

b i Large size, internal membranes

   ii Two of: presence of chloroplasts, cell wall and large vacuole.

c The cell membrane is the structure involved. The concentration is maintained through active transport. This is carried out by membrane carrier proteins coupled to a source of energy. The carrier has binding sites that allow iodide ions to bind to the side of the membrane in sea water. They function in one direction only, like valves, and require energy to change shape and move the iodide ion across the membrane.

d

Name of stage	Site within a chloroplast
Light dependent	Thylakoid membranes
Light independent	Stroma

e The rate of photosynthesis could be increased by two of:

- increasing the amount of sunlight available
- ensuring that each plant had plenty of water available
- artificially supplying carbon dioxide into the greenhouse
- ensuring the temperature is maintained at a warm level, to assist enzyme function.

2 a Tissue A

b Any of: transpiration, cohesion of water molecules, adhesion of water molecules

c Tissue B carries the glucose produced by photosynthesis out of the leaf to where it is needed in the plant.

3 a The highest concentration of ADH (4.7 pg/mL) would most likely be found in a desert animal. The urine output and sweating are low because reabsorption of water has increased. This conserves water.

b ADH levels would decrease because reabsorption of water needs to decrease.

9780170411660

c

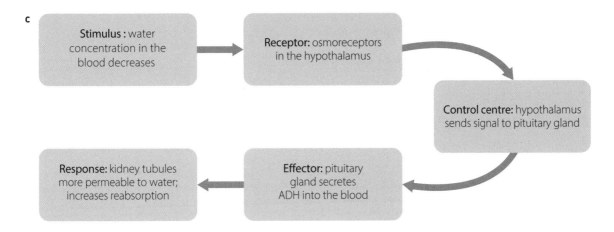

Stimulus : water concentration in the blood decreases → Receptor: osmoreceptors in the hypothalamus → Control centre: hypothalamus sends signal to pituitary gland → Effector: pituitary gland secretes ADH into the blood → Response: kidney tubules more permeable to water; increases reabsorption

d Any two of the following.

Nervous system	Endocrine system
Uses an electrochemical signal	Uses a chemical signal
Short duration	Long duration
Works quickly (milliseconds)	Works slowly (minutes/hours)
Circulated via neurons	Circulated via bloodstream
Distribution is general	Distribution is specific

4 a Through nasal secretions spread through sneezing, contamination of food and water in troughs.

b Three of: direct contact, contact with body fluids, contaminated food, through water and disease-specific vectors.

c The vaccine stimulates an immune response stimulating B lymphocyte cells to divide rapidly into B-plasma cells and B-memory cells. Antibodies secreted from the B-plasma cells bind to the *Streptococcus equi* bacteria.

Repeated doses of the vaccine promote a secondary response that results in a faster and greater production of antibody.

d Passive immunity occurs when antibodies are given. This provides protection from *Streptococcus equi*, but only for as long as those antibodies last. Because there are no memory cells stimulated by a vaccine (active immunity), the horse will not be immune if they encounter the pathogen in the future.

e Vaccination if available, quarantine of infected people, improved hygiene such as washing hands and cleanliness around food preparation, cook food.